T0189477

Studies in Computational Intelligence

Volume 791

Series editor

Janusz Kacprzyk, Polish Academy of Sciences, Warsaw, Poland
e-mail: kacprzyk@ibspan.waw.pl

The series "Studies in Computational Intelligence" (SCI) publishes new developments and advances in the various areas of computational intelligence-quickly and with a high quality. The intent is to cover the theory, applications, and design methods of computational intelligence, as embedded in the fields of engineering, computer science, physics and life sciences, as well as the methodologies behind them. The series contains monographs, lecture notes and edited volumes in computational intelligence spanning the areas of neural networks, connectionist systems, genetic algorithms, evolutionary computation, artificial intelligence, cellular automata, self-organizing systems, soft computing, fuzzy systems, and hybrid intelligent systems. Of particular value to both the contributors and the readership are the short publication timeframe and the world-wide distribution, which enable both wide and rapid dissemination of research output.

More information about this series at http://www.springer.com/series/7092

Roger Lee

Editor

Computer and Information Science

 Springer

Editor
Roger Lee
Software Engineering and Information
  Technology Institute
Central Michigan University
Mt. Pleasant, MI, USA

ISSN 1860-949X   ISSN 1860-9503 (electronic)
Studies in Computational Intelligence
ISBN 978-3-030-07521-7   ISBN 978-3-319-98693-7 (eBook)
https://doi.org/10.1007/978-3-319-98693-7

© Springer Nature Switzerland AG 2019
Softcover re-print of the Hardcover 1st edition 2019
This work is subject to copyright. All rights are reserved by the Publisher, whether the whole or part
of the material is concerned, specifically the rights of translation, reprinting, reuse of illustrations,
recitation, broadcasting, reproduction on microfilms or in any other physical way, and transmission
or information storage and retrieval, electronic adaptation, computer software, or by similar or dissimilar
methodology now known or hereafter developed.
The use of general descriptive names, registered names, trademarks, service marks, etc. in this
publication does not imply, even in the absence of a specific statement, that such names are exempt from
the relevant protective laws and regulations and therefore free for general use.
The publisher, the authors, and the editors are safe to assume that the advice and information in this
book are believed to be true and accurate at the date of publication. Neither the publisher nor the
authors or the editors give a warranty, express or implied, with respect to the material contained herein or
for any errors or omissions that may have been made. The publisher remains neutral with regard to
jurisdictional claims in published maps and institutional affiliations.

This Springer imprint is published by the registered company Springer Nature Switzerland AG
The registered company address is: Gewerbestrasse 11, 6330 Cham, Switzerland

Foreword

The purpose of the 17th IEEE/ACIS International Conference on Computer and Information Science (ICIS 2018) held on June 6–8, 2018 in Singapore was to together researchers, scientists, engineers, industry practitioners, and students to discuss, encourage, and exchange new ideas, research results, and experiences on all aspects of Applied Computers and Information Technology, and to discuss the practical challenges encountered along the way and the solutions adopted to solve them. The conference organizers have selected the best 13 papers from those papers accepted for presentation at the conference in order to publish them in this volume. The papers were chosen based on review scores submitted by members of the program committee and underwent further rigorous rounds of review.

In Chapter "The Analysis of the Appropriateness of Information Education Curriculum Standard Model for Elementary School in Korea", Namje Park, Younghoon Sung, Youngsik Jeong, Soo-Bum Shin, and Chul Kim conducted a survey among elementary school teachers to select key concepts necessary for information education and relevant achievement criteria were suggested for the Appropriateness of Information Education Curriculum Standard Model for Elementary School in Korea.

In Chapter "New Gene Selection Method Using Gene Expression Programing Approach on Microarray Data Sets", Russul Alanni, Jingyu Hou, Hasseeb Azzawi, and Yong Xiang propose a new feature selection method for microarray data sets. The method consists of the Gain Ratio (GR) and Improved Gene Expression Programming (IGEP) algorithms which are for gene filtering and feature selection, respectively.

In Chapter "A Novel Differential Selection Method Based on Singular Value Decomposition Entropy for Solving Real-World Problems", Rashmi and Udayan Ghose present an optimal feature selection approach based on Differential Evolution (DE) and SVD entropy is proposed. The functioning of the proposed approach is examined on available UCI data sets. This approach provides ranked features by optimizing SVD entropy using the DE.

In Chapter "Power Consumption Aware Machine Learning Attack for Feed-Forward Arbiter PUF", Yusuke Nozaki and Masaya Yoshikawa propose a new machine learning attack using power consumption waveforms for the feed-forward arbiter PUF which is one of the typical PUFs. In experiments on a Field-Programmable Gate Array (FPGA), the validity of the proposed analysis method and the vulnerability of the feed-forward arbiter PUF were clarified.

In Chapter "Adaptive Generative Initialization in Transfer Learning", Wenjun Bai, Changqin Quan, and Zhi-Wei Luo propose a novel initialization technique, i.e., adapted generative initialization. Not limit to boost the task transfer, more importantly, the proposed initialization can also bound the transfer benefits in defending the devastating negative transfer.

In Chapter "A Multicriteria Group Decision Making Approach for Evaluating Renewable Power Generation Sources", Santoso Wibowo and Srimannarayana Grandhi formulate the renewable power generation sources' performance evaluation problem as a multicriteria group decision making problem, and present a new multicriteria group decision making approach for effectively evaluating the performance of renewable power generation sources.

In Chapter "Incremental Singular Value Decomposition Using Extended Power Method", Sharad Gupta and Sudip Sanyal present a novel method to perform Incremental Singular Value Decomposition (ISVD) using an adaptation of the power method for diagonalization of matrices. They find that the efficiency of the procedure depends on the initial values and present two alternative ways for initialization.

In Chapter "Personalized Landmark Recommendation for Language-Specific Users by Open Data Mining", Siya Bao, Masao Yanagisawa, and Nozomu Togawa propose a personalized landmark recommendation algorithm aiming at exploring new sights into the determinants of landmark satisfaction prediction.

In Joint Radio Resource Allocation in LTE-A Relay Networks with Carrier Aggregation", Jichiang Tsai introduces an efficient greedy scheduling algorithm to perform joint downlink radio resource allocation in an LTE-A outband relay network, additionally taking the above issues into consideration. Our proposed technique can also apply to the upcoming 5G system.

In Chapter "Efficient Feature Points for Myanmar Traffic Sign Recognition", Kay Thinzar Phu and Lwin Lwin Oo propose an efficient feature point for Traffic Sign Recognition (TSR) system. This system composed of adaptive thresholding method based on RGB color detection, shape validation, feature extraction, and Adaptive Neuro-Fuzzy Inference System (ANFIS).

In Chapter "A Study on the Operation of National Defense Strategic Operation System Using Drones", Won-Seok Song, Si-Young Lee, Bum-Taek Lim, Eun-tack Im, and Gwang Yong Gim use "Electronic Warfare Subdivisions" of Joint Chiefs of Staff in reviewing priority in preparation for EW system to mount on drones and examine the difference in priority by EW sector. In this paper, they will explain the concept and prospect of drones and EW system and propose a method of applying a relative weight and calculating importance (priority) by EW sector in order to enhance the capability of future EW.

In Chapter "A Mechanism for Online and Dynamic Forecasting of Monthly Electric Load Consumption Using Parallel Adaptive Multilayer Perceptron (PAMLP)", J. T. Lalis, B. D. Gerardo, and Yungcheol Byun present a study that is based on time series modeling with special application to electric load consumptions modeling in a distributed environment. Existing theories and applications about artificial neural networks, backpropagation learning method, Nguyen-Widrow weight initialization technique, and autocorrelation analysis were explored and applied. An adaptive stopping criterion algorithm was also integrated to BP to enable the ANN to converge on the global minimum and stop the training process without human intervention.

In Chapter "An Influence on Online Entrepreneurship Education Platform Utilization and Self-deterministic to College Students' Entrepreneurial Intention", Sungtaek Lee, Hoo-Ki Lee, Hong-Seok Ki, and Gwang Yong Gim discuss the establish entrepreneurship education at universities as a system that experts in the relevant field and conducts in a separate field from the major education. However, there is also a gap in the entrepreneurship education depending on the competence of the experts for entrepreneurship education. In order to solve such gaps in education, online education platforms such as MOOC are expected to be needed in entrepreneurship education.

It is our sincere hope that this volume provides stimulation and inspiration, and that it will be used as a foundation for works to come.

Singapore, Singapore   Wei Xiong
                                                           Institute for Infocomm Research

Sault Ste. Marie, Canada   Simon Xu
June 2018   Algoma University

Contents

**The Analysis of the Appropriateness of Information Education
Curriculum Standard Model for Elementary School in Korea** 1
Namje Park, Younghoon Sung, Youngsik Jeong, Soo-Bum Shin
and Chul Kim

**New Gene Selection Method Using Gene Expression Programing
Approach on Microarray Data Sets** 17
Russul Alanni, Jingyu Hou, Hasseeb Azzawi and Yong Xiang

**A Novel Differential Selection Method Based on Singular Value
Decomposition Entropy for Solving Real-World Problems** 33
Rashmi and Udayan Ghose

**Power Consumption Aware Machine Learning Attack
for Feed-Forward Arbiter PUF** 49
Yusuke Nozaki and Masaya Yoshikawa

Adaptive Generative Initialization in Transfer Learning 63
Wenjun Bai, Changqin Quan and Zhi-Wei Luo

**A Multicriteria Group Decision Making Approach for Evaluating
Renewable Power Generation Sources** 75
Santoso Wibowo and Srimannarayana Grandhi

**Incremental Singular Value Decomposition Using Extended
Power Method** 87
Sharad Gupta and Sudip Sanyal

**Personalized Landmark Recommendation for Language-Specific
Users by Open Data Mining** 107
Siya Bao, Masao Yanagisawa and Nozomu Togawa

**Joint Radio Resource Allocation in LTE-A Relay Networks
with Carrier Aggregation** 123
Jichiang Tsai

Efficient Feature Points for Myanmar Traffic Sign Recognition 141
Kay Thinzar Phu and Lwin Lwin Oo

**A Study on the Operation of National Defense Strategic Operation
System Using Drones** 155
Won-Seok Song, Si-Young Lee, Bum-Taek Lim, Eun-tack Im
and Gwang Yong Gim

**A Mechanism for Online and Dynamic Forecasting of Monthly
Electric Load Consumption Using Parallel Adaptive Multilayer
Perceptron (PAMLP)** 177
J. T. Lalis, B. D. Gerardo and Yungcheol Byun

**An Influence on Online Entrepreneurship Education Platform
Utilization and Self-deterministic to College Students' Entrepreneurial
Intention** 189
Sung Taek Lee, Hoo Ki Lee, Hong Seok Ki and Gwang Yong Gim

Author Index 213

Contributors

Russul Alanni School of Information Technology, Deakin University, Geelong, VIC, Australia

Hasseeb Azzawi School of Information Technology, Deakin University, Geelong, VIC, Australia

Wenjun Bai Department of Computational Science, Kobe University, Nada, Kobe, Japan

Siya Bao Department of Computer Science and Communications Engineering, Waseda University, Tokyo, Japan

Yungcheol Byun Department of Computer Engineering, Jeju National University, Jeju, Korea

B. D. Gerardo Institute of ICT, West Visayas State University, Lapaz Iloilo City, Philippines

Udayan Ghose University School of Information Communication and Technology, Guru Gobind Singh Indraprastha University, Dwarka, New Delhi, India

Gwang Yong Gim Department of Business Administration, Department of IT Policy and Management, Soongsil University, Seoul, Republic of Korea

Srimannarayana Grandhi School of Engineering & Technology, CQUniversity, Melbourne, Australia

Sharad Gupta Information Technology, Indian Institute of Information Technology, Allahabad, Allahabad, India

Jingyu Hou School of Information Technology, Deakin University, Geelong, VIC, Australia

Eun-tack Im Department of Business Administration, Soongsil University, Seoul, Republic of Korea

Youngsik Jeong Department of Computer Education, Jeonju National University of Education, Jeonju, South Korea

Hong Seok Ki Department of IT Policy and Management, Soongsil University, Seoul, Republic of Korea

Chul Kim Department of Computer Education, Gwangju National University of Education, Gwangju, South Korea

J. T. Lalis College of Computer Studies, La Salle University, Ozamiz City, Philippines

Hoo Ki Lee Department of IT Policy and Management, Soongsil University, Seoul, Republic of Korea

Si-Young Lee Department of IT Policy and Management, Soongsil University, Seoul, Republic of Korea

Sung Taek Lee Department of IT Policy and Management, Soongsil University, Seoul, Republic of Korea

Bum-Taek Lim Department of IT Policy and Management, Soongsil University, Seoul, Republic of Korea

Zhi-Wei Luo Department of Computational Science, Kobe University, Nada, Kobe, Japan

Yusuke Nozaki Department of Information Engineering, Meijo University, Tenpaku-ku, Nagoya, Aichi, Japan

Lwin Lwin Oo Department of Natural Science, University of Computer Studies, Mandalay, Myanmar

Namje Park Department of Computer Education, Teachers College, Jeju National University, Jeju, South Korea

Kay Thinzar Phu University of Computer Studies, Mandalay, Myanmar

Changqin Quan Department of Computational Science, Kobe University, Nada, Kobe, Japan

Rashmi University School of Information Communication and Technology, Amity School of Engineering and Technology, Guru Gobind Singh Indraprastha University, Dwarka, New Delhi, India

Sudip Sanyal Computer Science and Engineering, BML Munjal University, Gurgaon, India

Soo-Bum Shin Department of Computer Education, Gongju National University of Education, Gongju, South Korea

Won-Seok Song Department of IT Policy and Management, Soongsil University, Seoul, Republic of Korea

Younghoon Sung Department of Computer Education, Chinju National University of Education, Jinju, South Korea

Nozomu Togawa Department of Computer Science and Communications Engineering, Waseda University, Tokyo, Japan

Jichiang Tsai Department of Electrical Engineering, National Chung Hsing University, Taichung, Taiwan, ROC

Santoso Wibowo School of Engineering & Technology, CQUniversity, Melbourne, Australia

Yong Xiang School of Information Technology, Deakin University, Geelong, VIC, Australia

Masao Yanagisawa Department of Computer Science and Communications Engineering, Waseda University, Tokyo, Japan

Masaya Yoshikawa Department of Information Engineering, Meijo University, Tenpaku-ku, Nagoya, Aichi, Japan

The Analysis of the Appropriateness of Information Education Curriculum Standard Model for Elementary School in Korea

Namje Park, Younghoon Sung, Youngsik Jeong, Soo-Bum Shin and Chul Kim

Abstract In Korea, it has been recently claimed that SW education hours included in the 2015 curriculum are insufficient and more hours are being requested to be allocated to SW education. Under the circumstances, the Korean Association of Information Education (hereinafter "KAIE") has developed standard models of information education curriculum in 2014–2016, targeting the revision of curriculum for 2020. However, the KAIE's curriculum is not well associated with information education curriculum for middle and high school students and is not based on key concept-centered achievements, which has made teachers reorganize the curriculum and accordingly suffer difficulties to operate. In this paper, Delphi survey was conducted among elementary school teachers to select key concepts necessary for information education and relevant achievement criteria were suggested.

Keywords Information education · Education curriculum
Elementary School Korea Education

N. Park
Department of Computer Education, Teachers College, Jeju National University, Jeju,
South Korea
e-mail: namjepark@jejunu.ac.kr

Y. Sung (✉)
Department of Computer Education, Chinju National University of Education, Jinju,
South Korea
e-mail: yhsung@cue.ac.kr

Y. Jeong
Department of Computer Education, Jeonju National University of Education, Jeonju, South Korea

S.-B. Shin
Department of Computer Education, Gongju National University of Education, Gongju, South
Korea
e-mail: ssb@gjue.ac.kr

C. Kim
Department of Computer Education, Gwangju National University of Education, Gwangju, South
Korea
e-mail: chkim@gnue.ac.kr

© Springer Nature Switzerland AG 2019
R. Lee (ed.), *Computer and Information Science*, Studies in Computational
Intelligence 791, https://doi.org/10.1007/978-3-319-98693-7_1

1   Introduction

In November 2017, the Korean government announced plans for the Fourth Industrial Revolution and added that it would increase software (hereinafter "SW") education to nurture outstanding individuals with creative thinking and troubleshooting ability and provide SW-teaching professionals by re-educating the current faculty to strengthen their SW educational capabilities and running an SW curriculum re-search association. Before this announcement, the Ministry of Education of Korea had included SW education in the Practical Course curriculum revised in 2015 [1–3]. In other words, it emphasized SW education in elementary schools for 2019 onward, including understanding of SW, procedural troubleshooting, and programming elements and architecture for 5th to 6th graders in elementary school [1, 4–7]. The ministry also distributed operating guidelines for SW education to introduce SW subject before the 2015 revised curriculum is actually applied to the education field. However, 17-h SW education is insufficient in reality com-pared to cases of other countries such as the US, the UK, and India, and it is required to include an information subject in elementary school curriculum in preparation for the next curriculum [2, 8].

KAIE has developed 'standard models of information education curriculum' for 2015 revised curriculum to provide information-subject education early for elementary school students. Starting by the development of information education curriculum for elementary school in 2014, KAIE has developed a SW education curriculum model in 2015 and an information education curriculum model in 2016 with revisions [4–7]. In 2016, KAIE conducted a Delphi survey among professors in the Department of Computer Education at universities of education across the country including Jeju National University, and based on the survey results, it derived agreement on main terminology, education scope, and education contents of the information-subject curriculum [9, 10]. However, the standard model developed by KAIE focuses on elementary school and does not encompass curriculums for elementary, middle and high school students, and it is not closely associated with the 2015 revised information-subject curriculum. In addition, to be proposed as a curriculum model in preparation for the next-term curriculum, the model should have been in accordance with the 2015 revised curriculum system, but it is composed of characteristic, objectives, and achievement criteria.

In this regard, this research reorganized the information education curriculum for elementary school students with key concepts introduced in the 2015 revised curriculum, while considering information-subject curriculums for middle and high school students.

2 Creating the Key Concept-Centered Information-Education Curriculum

To help the next-term curriculum revision in the condition where there is no information education curriculum for elementary schools in Korea, it is required to understand the direction of the 2015 revised curriculum to propose an information education curriculum model [1]. Therefore, this research examined the elements and important concepts of the 2015 revised curriculum.

2.1 Elements of the 2015 Revised Curriculum in Korea

The 2015 revised curriculum roughly consists of four elements: characteristics of subjects, objectives, contents and achievement criteria, and teaching method and evaluation direction. First, the Characteristics of Subjects part describes the necessity and role of education by subject, scope of subjects, and important capabilities developed by each subject. This part outlines the unique characteristics of each subject and suggests relevance between human character and educational objective. Furthermore, characteristics of each subject by educational level, connection, and relation with other subjects are described. Second, the Objective part presents curriculum direction and learning destination by considering relevance between human character and educational objective. Objectives include overall, per-educational level objective, and detailed objectives, and objectives per grade can be suggested. As possible, the objectives shall include key capabilities and in describing detailed objectives, active verbs are used along with three to four sub items. Third, the Contents and Achievement Criteria part presents a subject content table to show the scope, key concepts, content (generalized knowledge), and functions by grade group, and their association. In addition, this part proposes functions that are expected to be performed by students through the subject and details contents that show the association between grades rather than presenting materials and themes of each subject. By integrating the functions and contents, achievement criteria are stated in sentence. Fourth, the Teaching Method and Evaluation Direction part describes key capabilities and suggests teaching methods and evaluation plans taking into account the content and scope of each subject, relevance and connection between subjects, association with careers, and trans-subject learning subjects.

2.2 Key Concepts of the 2015 Revised Curriculum in Korea

In the 2009 revised curriculum, the content of each subject was presented without key points of the composition of the content, and it was simply divided by area, which had a limitation to show the degree of association, range, and integration. In other words, how specific contents are selected and organized is not sufficiently explained and if it is presented based on a theme, it is difficult for teachers to reorganize [1]. To solve the problem, the 2015 revised curriculum included areas, key concept, generalized knowledge, content elements, and functions as the system of contents.

First, areas mean the top-level rule or system that best reveals characteristics of subjects and organizes the learning contents of subjects.

Second, key concepts mean the most basic concept or principles which a subject is based on, and key points that shall be learned by the subject even through detailed fact or information is forgotten.

Third, generalized knowledge means knowledge that students should have as they study in class and school and general principles of learning contents for all grades.

Fourth, content elements mean important, connotative, and essential contents that should be learned by grade and school grade, based on generalized knowledge.

Fifth, functions mean subject-specific exploration process and thinking function that should be conducted or that are expected to be conducted based on "content (knowledge)." In particular, key concepts introduced in the 2015 revised curriculum are must-know concepts after students learn each subject and as the concepts are associated with other subjects and helpful for solving practical problems, teachers shall identify relevance between content elements based on key concepts when reorganizing curriculum, and pursue integration in a subject and associated learning between subjects.

2.3   Consideration of Information Education Curriculum of Elementary School in Korea

To establish a basic contents system of information education curriculum in Korea, this research analyzed computer education-related curriculums in Korea and the objectives, composition of each area, and main characteristics of KAIE curriculum, as follows:

As for the information-related curriculum in Korea, operating guidelines for ICT Education implemented 1-h discretionary activity for ICT utilization in 2000, and the educational content focused on SW utilization [7, 9, 11]. However, in the 2007 revised curriculum, creative experience activities were reduced and accordingly the existing 1-h ICT education was in fact abolished [9]. According to PISA ICT statistical data from surveys on learners' capabilities of digital literacy conducted in 2017 among students in OECD countries, Korean students was revealed to be in the worst level in ICT utilization, out of 31 countries [9, 12]. To solve the problem, the Korean government should pursue basic quality education on digital literacy, which has been already implemented in advanced countries including the US and the UK and strengthen contents system of curriculum so as to learn basic concepts and principles on computer science [13, 14].

Table 1 Revision History of KAIE's Study of Information Education Curriculum Korea

Division	2014	2015	2016
Purpose	• Information-subject content system for elementary school	• SW curriculum model	• Standard model of information-subject curriculum
Areas	• Computer system • SW production • Convergence activities	• Software • Computer system • Convergence activities	• Software • Computing system • Information living • Computing thinking skills
Sub-areas	• 44 detailed areas	• 10 areas • 140 achievement criteria	• 8 areas • 162 achievement criteria
Target students	• 1st graders at elementary school to 3rd graders at middle school	• Preschoolers to 6th graders at elementary school	• Elementary school students to high school students
Main characters	• Consists of areas and sub-areas • Presents name of information science subjects • Distinguished by school-grade (elementary school/middle school) • Information life (recognition of a prob-lem) → computer system (principle learning) → SW production (production) • Integrated into the information life convergence activity area	• Areas, sub-areas, and achievement criteria • Step-by-step composition regardless of grade • Strengthens SW education • Reorganizes 10 areas through improvement of detailed areas • 2 achievement criteria for each of 7 steps and by area	• Establishes content system (Upper-level area–lower-level area—achievement criteria) • Convergence of the upper-level area and the computing thinking skills • Reorganizes into 8 areas through improvement of detailed areas • 3 steps for elementary school, 1 step for middle school, and 1 step for high school (Total 5 steps)

2.4   Composition of KAIE Curriculum in Korea

Revision history of KAIE curriculum from 2014 to 2016 is as follows [4–7, 15] (Table 1).

First, upper-level areas have been largely changed. In 2014, it consisted of computer system, software production, and convergence activities. In 2015, software production was changed to software and in 2016, convergence activities were divided into information living and computing thinking skills. In particular, computing thinking skills were presented in a form of associating with relevant content elements of each area.

Second, in 2014, there were a total of five levels: first to third levels for elementary school, fourth level for middle school, and fifth level for high school [4, 15]. However, in 2015, seven levels were presented without division of grades so that students can move across levels according to their achievement [5]. In 2016, the system was back to 5 levels in line with elementary and middle school curriculum [6].

Third, information education curriculum was set to be introduced from the first grade at elementary school [7, 16]. Operating guidelines for 2000 ICT education set an hour of ICT education out of two hours of discretionary activities as mandatory and time for ICT education was secured, but since the 2007 revised curriculum, discretionary activity time has been reduced to one hour, and therefore, it is hard to have time for ICT education [7, 11]. In addition, time for ICT education is not secured in the operating guidelines for 2015 SW Education, and ICT education is provided only in some SW education leading schools or research schools [17].

2.5 Overseas Trend

Recent trend of computer science-related curriculum in advanced countries such as the U.S.A. and the UK is as follows:

The US K12 Computer Science Framework Steering Committee (2017) materials propose that there should be content elements classified into concepts and activities for practical understanding of and approach to computer science and the elements were fused as if blocks were connected, so that students learn the contents by associating with achievement criteria [2]. In addition, ideas with which abstract and general elements presented by key concepts are associated with other concepts to apply are provided and they are utilized as an element to enrich the key concept [2]. The US ISTE has changed its objectives for computer education for capable learners from learning to use technology in 1998 to learning with technology in 2007 and it pursued transformative learning using technology in 2016 [18]. These changes were made to emphasize activities considering learners' level, which is similar to computer education curriculum of India emphasizing convergence between knowledge elements, learning concepts and exploration methods [4, 17, 18].

As shown in Fig. 1, detailed trend of information education curriculum researched and developed by KAIE is as follows [4–7, 9, 15].

First, purpose of composition of curriculum was started by the establishment of information-subject contents system required for elementary school, through detailed SW curriculum model, and developed to a standard model for information education curriculum that can comprehensively apply to elementary school to high school.

Second, areas composing the curriculum have been changed according to demand and trend, from computer system, SW production, and convergence activities in 2014 to SW, computer system, and convergence activities with the 2015 SW curriculum

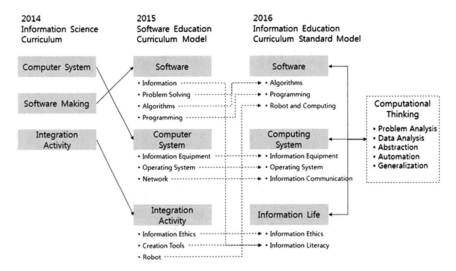

Fig. 1 KAIE information education curriculum (2014–2016) [9]

was introduced. As importance of computing thinking skills became grater in 2016, contents system consisting of SW, computing system, information living, and computing thinking skills was developed [9].

Third, as for the computing system area, the information device, operating system, network areas were included, out of which network area was expanded to the information search area through information device in everyday life, and ICT area focusing on communication and expression based on information search [15].

3 Proposed Information Education Curriculum Framework for Elementary School

Key areas of information education curriculum proposed by this paper are divided into four areas as follow:

Even though those are four areas, learning content includes computing system, software, and information and culture, and the information thinking skills area encompasses the before-mentioned three areas, as shown in Table 2. This structure in which the computational thinking skills area embraces the rest three areas expresses that in the curriculum, the computing system, information and culture, and software areas are all associated with computational thinking skills. Detailed information education curriculum in elementary school and content system are as follows.

Content direction by Information-subject content system area is as follows [10].

Table 2 Information education curriculum framework for elementary school [10]

Area			Step 1		Step 2	
			3rd Grade	4th Grade	5th Grade	6th Grade
Software	Algorithm (4)	– Concept – Expression – Processing – Utilization	– Understand the work procedure – Express the work procedure – Analyze the meaning of the problem – Relations between algorithm and program	– Meaning of algorithm – Express the problem-solving process with words and writing – Express a solution considering conditions – Experience algorithm	– Relations between algorithm and literacy – Express problem solving process in figure and symbol – Flow chart and pseudo-code – Explain current state and target state of the problem	– Operating principles of algorithm – Express pattern finding and problem solving – Express various flow charts – Improve algorithm
	Programing (4)	– Concept – Expression – Processing – Utilization	– Meaning of program – Relations between algorithm and program – Program operation – Simple block programing	– Necessity and role of programing language – Understanding and utilization of parameters – Understanding of arithmetic operator – Develop simple algorithm	– Types of programing language – Necessity and use of loop – Understand comparison operator and logic operator – Text programing language	– Explore program operating principles – Control structure – Understanding of program debugging – Improvement of program
	Robot and computing (4)	– Concept – Expression – Processing – Utilization	– Definition of robot – Types and composition of robot – Rules for robot and safe use – Understanding of robot movements	– Operating principles of robot – Understand rotational movements – Produce simple movement robot – Explain simple movement robot	– Experience robot and problem solving – Understand sensor and various operations – Produce various robot operating works – Explain robot operating works	– Understand robots in everyday life – Manufacture simple sensor robot – Explain simple sensor robot – Design rules and apply and utilize them

(continued)

Table 2 (continued)

Area			Step 1		Step 2	
			3rd Grade	4th Grade	5th Grade	6th Grade
Computing system	Information devices (4)	– Concept – Structure – Resources – Processing	– Learn about information devices in everyday life – Distinguish information devices – Connect input device of information device – Roles of information device	– Roles of information devices in everyday life – Structure of various information devices – Connect output device for information device – Use input and output devices	– Information device and SW – Structure of operation device of information device – Connect identification device of information device – Utilize of information identification device	– Information device and programing – Information device and IoT structure – Connect complex device for information device – Utilize complex device
	Operating system (4)	– Concept – Structure – Resources – Processing	– Understanding of work management – Computer system architecture – Create and copy files – Protection of computer	– Work management function – Examines various operating systems – File attributes and compression – Computer state management	– Concept of operating system – Role of operating system – File system management – Computer account management	– Computer system – Operating system operation – Program management – Computer resource management
	Information and communications (4)	– Concept – Structure – Resources – Processing	– Understanding of ICT in everyday life – Observe information transferring tools – Communications method and information transfer – Types of information transfer signals	– Roles of information devices in everyday life – Observe ICT devices – Transfer information through ICT device – Process of information transfer through ICT device	– Understanding of Internet connection – Characteristics of Internet access devices – Internet connection – Utilize information transfer program	– Understanding of internet and IoT – IoT devices – Wired and wireless Internet connection – Internet in everyday life

(continued)

Table 2 (continued)

Area		Step 1		Step 2	
		3rd Grade	4th Grade	5th Grade	6th Grade
Information and culture	Information ethics (4) – Internet etiquette – Protection of personal Information – Protection of copyright	– Internet etiquette in language use – Protect my personal information – Light and shadow of the Internet – How to use works	– Internet etiquette and practice – Protect other's personal information – Cyber bullying and countermeasures – How to mark a source of a work	– SNS etiquette – Personal information infringement and countermeasures – Prevention of and countermeasures against overuse of Internet – Creative Commons License	– AI and ethics – Internet account and password – Harmful contents and blocking methods – Copyright infringement cases
	Information utilization (6) – Preparation of document – Multimedia Edit – Document Sharing – Communication	– Prepare and modify of document (basic) – View multimedia data – Prepare a document with figures – Internet search – SNS search – Use email	– Prepare and share documents in collaboration – Prepare digital note – Prepare presentation documents – Photographing and videotaping – Sound recording – Share information and opinions	– Prepare basic document with formula – Organize document for presentation – Sound edit – Image edit – Share document between devices – Access web app and install app	– Prepare online survey document and analyze – Prepare document with formulas in collaboration – Edit and conversion of video and experience of AR/VR – Internet search operator – Cloud information management

3.1 Software Area

The algorithm, programing, and robot and computing areas composing the SW area are required to have content system with association with other knowledge considered, by connecting the sub-areas under other upper-level areas.

Since SW is traditionally in a realm of computer science, it should belong to the computing system area of information-subject curriculum, but on the other hand, as it is a key area and influence of SW education shall be reflected, it should be a separate area. In addition, the algorithm area was separated to emphasize data-driven practical activities as well as independent problem solving ability and computational thinking skills, apart from programing. In the intermediate level of programming, specific programing language was not emphasized so that programing language can be selected and utilized in accordance with the conditions of the school.

3.2 Computing System Area

The computing system area was the computer system area in 2014 and 2015, but as information processing and the meaning of communication are emphasized in 2016, it was changed to computing system. Initially, the scope of information components comprising a computer included main computer, peripheral devices, and network, but it has expanded to operating principles and theory learning of HW and SW used in everyday life, as technology has changed and advanced. Accordingly, changes have been made in sub-areas. For example, information devices and operating system have improved in terms of achievement criteria and the network element was revised to the ICT area, changing from passive interpretation of meaning centered on information transfer to information search and communication and expression using information devices. Contents system of the information devices, operating system and ICT, which are sub-areas of the computing system, was composed by considering association with knowledge of other lower-level areas. With regard to concepts and exploration method of computer science, the content was designed to support learners to process and express information as they intends, based on the viewpoint of computing thinking skills created in the course of learning activities of both theory and practice.

3.3 Information Culture Area

To increase association with information education curriculum of elementary school and middle school, name of the upper-level area proposed by KAIE was re-vised from information living to information culture. As the information culture area is

$$CVR = \frac{N_e - \frac{N}{2}}{\frac{N}{2}} \qquad \text{Agreement} = 1 - \frac{Q_3 - Q_1}{Mdn}$$

Fig. 2 CVR for verification [10, 18]. (*Ne: Number of panels who respond "valid", *N: Number of panels participating in the research, *Q_1, Q_3: Coefficient of first quarter and third quarter, *Mdn: Median value)

closely related to living space of students, the content was proposed to gradually expand living space taking into account the development stage of students.

Information living, the existing upper-level area, was changed to information culture and it was divided into two sub-areas: information ethics and information utilization. As for content elements, there were total 17 elements: 4 for 3rd, 4th, and 5th graders each and 5 for 6th graders.

4 Verification of the Content of the Proposed Information Education Curriculum for Elementary

To evaluate appropriateness of content elements of information-subject curriculum, Delphi survey was conducted among professors of the Department of Computer Education at universities of education across the country on importance of education contents and appropriateness of education timing [10]. Importance of education contents were evaluated with Likert scale and "very appropriate" was five points and "not appropriate at all" was 1 point. Education timing could be selected among 1st and 2nd grades of elementary school, 3rd and 4th grades of elementary school, 5th and 6th grades of elementary school, middle school, and high school. As for importance of education content for contents system, Content Validity Ratio (CVR) was calculated to analyze validity of survey results and when considering that the number of responded experts was 39, the results were deemed valid if CVR was 0.33 or greater (Fig. 2).

Analysis results by content system area are as follows. As for the SW area, the appropriateness of content composition and timing of applying achievement criteria was verified in the algorithm, programing, robot and computing areas [10, 19] (Table 3).

In the computing system area, the appropriateness of content composition and timing of applying achievement criteria was verified in the information device, operation system, information communication utilization areas [10, 20].

Main opinions of the respondents included opinions about content per grade for necessity of operating system management, that the level of understanding the meaning related to network information connection is somewhat high and causes confusion, and that methods of expressing wired and wireless Internet connection informa-

Table 3 SW area in curriculum [10]

Area	Content composition (Mean)	CVR	Consensus
Algorithm	4.57	0.84	0.85
Programing	4.52	0.81	0.85
Robot and computing	4.67	0.90	0.90
Total mean	4.58	0.85	0.87

Table 4 Computing system area in curriculum [10]

Area	Timing of applying achievement criteria (Mean)	CVR	Consensus
Algorithm	4.52	0.82	0.86
Programing	4.49	0.78	0.83
Robot and computing	4.58	0.84	0.89
Total mean	4.53	0.82	0.86

Table 5 Information culture area in curriculum [10]

Area	Content composition (Mean)	CVR	Consensus
Information ethics	4.77	0.95	0.81
Information utilization	4.38	0.81	0.80
Total mean	4.58	0.88	0.81
Area	*Timing of applying achievement criteria*	*CVR*	*Consensus*
Information ethics	4.70	0.91	0.95
Information utilization	4.63	0.88	0.84
Total mean	4.67	0.90	0.90

tion should be modified and supplemented to increase clarity of achievement criteria and enhance validity (Table 4).

In the information culture area, the appropriateness of content composition and timing of applying achievement criteria was verified in the information ethics and information utilization areas [10, 21, 22] (Table 5).

Many respondents said that since there is a lot of similar content between information ethics and information society, it is required to be integrated with a subject of cultivating civic awareness. Some respondents demanded that content of information ethics in preparation for the future society and now information ethics related to SNS utilization and AI robot should be added. Other respondents presented some requests on expressions and purification of terms according to the level of learners in the information utilization area [23–28].

5   Conclusion

The information education curriculum for elementary school students proposed in this paper consists of areas of software, computing system, and information culture, and it presented achievement criteria according to key concepts necessary to strengthen learners' active problem-solving ability, thereby establishing hierarchy and structure of the curriculum content elements. In addition, the curriculum was de-signed to be associated with other fields of study in the school education field and pursue collaboration between learners, knowledge convergence, and combination with elements in the computing thinking skills, through association with key concepts or converged associated concepts, thus fostering outstanding learners with creative capabilities who can solve a problem when faced with new challenge as well as strengthening problem-solving ability of learners.

Through this information education curriculum proposed in this research, further research is required to study various teaching system models that can be immediately applied to the real education field and evaluation strategies and methods to examine the entire curriculum.

Acknowledgements This work was supported by the Korea Foundation for the Advancement of Science & Creativity (KOFAC) grant funded by the Korean Government. And, this work was supported by the Ministry of Education of the Republic of Korea and the National Research Foundation of Korea (NRF-2017S1A5A2A01026664). The corresponding author is Younghoon Sung.

References

1. Ministry of Education: 2015 Revised Curriculum. Korea Ministry of Education (2015)
2. Kim, K.: An implications of computer education in Korea from the U.S., U.K. and Germany computer curriculums. J. Korean Assoc. Inf. Educ. **20**(4), 421–432 (2016)
3. Shin, S.-B., Kim, C., Park, N., Kim, K.-S., Sung, Y.-H., Jeong, Y.-S.: Convergence organization strategies of the computational thinking in informatics curriculums. J. Korean Assoc. Inf. Educ. **20**(6), 607–616 (2016)
4. Kim, K., Kim, C., Kim, H.-B., Jeong, I., Jeong, Y.-S., Ahn, S., Kim, C.W.: A study on contents of information science curriculum. J. Korean Assoc. Inf. Educ. **18**(1), 161–171 (2014)
5. Jeong, Y., Kim, K., Jeong, I., Kim, H., Kim, C., Jeongsu, Yu., Kim, C., Hong, M.: A development of the software education curriculum model for elementary students. J. Korean Assoc. Inf. Educ. **19**(4), 467–480 (2015)
6. Kim, C., Park, N., Sung, Y., Shin, S., Jeong, Y.: Development of information education curriculum standard model. KAIE Research Report (2016)
7. Jeong, Y., Kim, K., Jeong, I., Kim, H., Kim, C., Yu, J., Kim, C., Hong, M.: A development of the software education curriculum model for elementary students. J. Korean Assoc. Inf. Educ. 467–480 (2015)
8. Kim, J., Lee, W.: A study on India's CMC (Computer Masti Curriculum) based on Bruner's educational theories. J. Korean Assoc. Inf. Educ. **17**(6), 59–69 (2014)
9. Sung, Y., Park, N.: A study of the direction for developing KAIE computing system curriculum in elementary education. J. Korean Assoc. Inf. Educ. **21**(6), 701–710 (2017)
10. Kim, J., Seo, J., Kim, H., Lee, Y., Kim, C., Kim, D., Kim, S.: Policy Research Report on Software Education. Ministry of Science and ICT and KOFAC (2017)

11. Ministry of Education: 2005 Revised Information and Communication Technology Education Guidance. Korean Ministry of Education (2005)
12. Kim, K.: A study on ICT usability and availability of between Korean students and OECD students: focus on PISA 2015. J. Korean Assoc. Inf. Educ. **21**(3), 361–370 (2017)
13. Computing At School: Computing in the national curriculum: a guide for primary teachers. Computing at School (2013)
14. K12cs.org: K–12 Computer Science Framework. http://K12cs.org (2017)
15. Sung, Y., Park, N.: Development of contents structure for KAIE computing system area. Korean Assoc. Inf. Educ. Res. J. **8**(3), 9–14 (2017)
16. Ministry of Education: Information and Communication Technology Education Guidance. Korean Ministry of Education (2000)
17. Kim, H.: A development of curriculum model on information ethics and creation tools for elementary school students. J. Korean Assoc. Inf. Educ. **19**(4), 545–556 (2015)
18. Sung, Y., Park, N., Jeong, Y.: Development of algorithm and programming framework for information education curriculum standard model. J. Korean Assoc. Inf. Educ. **21**(1), 77–87 (2017)
19. Park, N.: The core competencies of SEL-based innovative creativity education. Int. J. Pure Appl. Math. **118**(19), 837–849 (2018)
20. Park, N.: STEAM education program: training program for financial engineering career. Int. J. Pure Appl. Math. **118**(19), 819–835 (2018)
21. Park, N., Kim, C., Shin, S.-B.: A study of information and communications framework for information education curriculum standard model. J. Korean Assoc. Inf. Educ. **21**(1), 127–136 (2017)
22. Park, N., Shin, S.-B., Kim, C.: The analysis of the appropriateness of the content standards of information, information appliances, and operating system in elementary school. J. Korean Assoc. Inf. Educ. **20**(6), 617–628 (2016)
23. Lee, D., Park, N.: Electronic identity information hiding methods using a secret sharing scheme in multimedia-centric internet of things environment. Pers. Ubiquitous Comput. 10.1007/s00779-017-1017-1 (2017)
24. Park, N., Bang, H.-C.: Mobile middleware platform for secure vessel traffic system in IoT service environment. Secur. Commun. Netw. **9**(6), 500–512 (2016)
25. Lee, D., Park, N.: Geocasting-based synchronization of Almanac on the maritime cloud for distributed smart surveillance. J. Supercomput. 1–16 (2016)
26. Park, N., Kim, M.: Implementation of load management application system using smart grid privacy policy in energy management service environment. Clust. Comput. **17**(3), 653–664 (2014)
27. Lee, D., Park, N.: A proposal of SH-Tree based data synchronization method for secure maritime cloud. J. Korea Inst. Inf. Secur. Cryptol. **26**(4), 929–940 (2016)
28. Park, N.: Implementation of inter-VTS data exchange format protocol based on mobile platform for next-generation vessel traffic service system. INFORMATION—Int. Interdiscip. J. **17**(10A), 4847–4856 (2014)

New Gene Selection Method Using Gene Expression Programing Approach on Microarray Data Sets

Russul Alanni, Jingyu Hou, Hasseeb Azzawi and Yong Xiang

Abstract Feature selection in machine learning and data mining facilitates the optimization of accuracy attained from the classifier with smallest number of features. The use of feature selection in microarray data mining is quite promising. However, usually it is hard to identify and select the feature genes from microarray data sets because multi-class categories and high dimensionality features exist in microarray data with a small-sized sample. Therefore, using good selection approaches to eliminate incomprehensibility and optimize prediction accuracy is becoming necessary, because it will help obtain genes that are relevant to sample classification when investigating large number of genes. In his paper, we propose a new feature selection method for microarray data sets. The method consists of the Gain Ratio (GR) and Improved Gene Expression Programming (IGEP) algorithms which are for gene filtering and feature selection respectively. Support Vector Machine (SVM) alongside with leave-one-out cross-validation (LOOCV) method was used to evaluate the proposed method on eight microarray datasets captured in the literature. The experimental results showed the effectiveness of the proposed method in selecting small number of features while generating higher classification accuracies compared with other existing feature selection approaches.

Keywords Feature selection · Gain ratio (GR) · Gene expression programming (GEP) · Support vector machine (SVM)

R. Alanni (✉) · J. Hou · H. Azzawi · Y. Xiang
School of Information Technology, Deakin University, Geelong, VIC, Australia
e-mail: ralanni@deaik.edu.au

J. Hou
e-mail: jingyu.hou@deaik.edu.au

H. Azzawi
e-mail: hazzawi@deaik.edu.au

Y. Xiang
e-mail: yong.xiang@deaik.edu.au

© Springer Nature Switzerland AG 2019
R. Lee (ed.), *Computer and Information Science*, Studies in Computational Intelligence 791, https://doi.org/10.1007/978-3-319-98693-7_2

1 Introduction

The application of DNA microarray technology makes it possible to have simultaneous monitory and measurement of multiple gene expression levels as they arise within one experiment. The microarray data classification desires creating efficient model capable of identifying the genes expressed variedly, thereby making prediction of class membership possible even for unknown samples. The use of microarray classification poses several challenges including the scarcity of samples relative to their high dimensionality alongside the occurrence of experimental variations when measuring the gene expression levels. This challenge is identified in data mining as a curse of dimensionality. It is associated with increasing the computational complexity when creating the model, especially when the dataset has many classes. Therefore, the process of classifying microarray data samples usually comprises feature selection and classifier design. The rational of feature selection is that a strong correlation only exists in a few gene expression data of specific phenotype among all genes investigated. Although there are thousands of genes, only a small portion of them have significant correlations. This makes feature selection methods essential to ensure the correct analysis of gene expression profiles for classification. However, it is a challenge to effectively select feature genes that are related to classification while still keeping a high prediction accuracy and eliminating errors.

There have been a number of methods that are capable of showing the informative cancer-related genes, reducing noise and spotting irrelevant genes [1–10]. However, gene selection is still a hot research area, and it needs to be improved to reduce the noise in the database and increase the classification performance. Variant hybrid methods based on Particle Swarm Optimization (PSO) have been proposed to solve the problems of gene selection. The Tabu Search (TS) has been embedded within the Binary Particle Swarm Optimization (BPSO) in [11] to prevent TS method from local optima problem. This method had achieved a satisfactory accuracy on average. However, to obtain that accuracy this method needs to select a large number of genes. Dashtban and Balafar [3] able to get a good accuracy with small number of selected genes. However, the number of selected genes still high compared with other method.

Recently developed Gene Expression Programming (GEP) [12], which is an evolutionary algorithm widely used in classification processes, applications and decision-making [13–19], has a potential to effectively select feature genes from microarray datasets. GEP has two main advantages: the flexibility which makes it easy to design an optimal model, and the power of achieving the target which is inspired by the ideas of genotype, phenotype and data visualisation. These advantages make it easy to simulate the dynamic process of achieving an optimal solution. However, GEP is not designed to solve the problems of gene selection and multi-classification. In order to make it applicable for our purposes, we replace the classification structure of GEP with support victor machine (SVM) algorithm and change the fitness function to a novel function we propose. The new fitness function takes into consideration the SVM based multi-classification and the number of selected genes.

SVM is incorporated into our GEP based method because it is a powerful classification algorithm which is widely used to solve the binary and multi-classification problems [20–24], and can achieve a high classification accuracy.

This study utilizes a two-stage approach in implementing feature selection. The first stage is to calculate the Gain Ratio (GR) values for all gene features. GR values are used to reduce noses from the microarray datasets. The second stage uses the Improved Gene Expression Programming (IGEP) to extract the smallest feature genes that are capable of optimizing the classification accuracy.

This paper is organised as follows. The proposed model is presented in Sect. 2. The experimental results are shown in Sect. 3 and finally we conclude our work and indicate the future research directions in Sect. 4.

2 Methods

2.1 Gain Ratio

Gain Ratio (GR) is a commonly used method for reducing the bias and noise attributes in a dataset [25]. This algorithm, when applied to a microarray dataset, ranks each gene in the microarray dataset based on the informative information the gene carries. A gene that carries more information has a higher rank. GR is an enhanced information gain algorithm because it takes the intrinsic information of genes into consideration. Intrinsic information incorporates the number of values a gene has. More information about information gain and intrinsic information can be found in [25–27]. The gain ratio of a gene g is defined as follows:

$$Gain\ Ratio\ (g) = \frac{Information\ Gain\ (g)}{Intrinsic\ Information\ (g)} \tag{1}$$

The WEKA 3.7.12 software [28] is used in this paper to calculate the GR values and rank the genes according to their GR values in a microarray dataset. The package of Gain Ratio is known as "GainRatioAttributeEval". The genes of each dataset are ranked by their GR values, and all the genes with GR value over zero are selected to be the input of the IGEP algorithm to find the best subset of the genes.

2.2 Gene Expression Programming

GEP is an evolutionary algorithm capable of emulating biological evolution on the basis of computer programming. Introduced in 2002 by Ferreira [12]. GEP inherits the advantages of the Genetic Algorithm (GA) [29] and the Genetic Programming (GP) [30] while overcoming their disadvantages. GEP consists of two parts. The first part is the characteristic linear chromosome (phenotype), which is inherited from GA but the chromosome in GA has only one gene while the chromosome in GEP can have more than one gene. A gene consists of head (h) and tail (t). The head contains functional elements such as (seq., +, −, ×, /) and terminal elements, while the tail contains terminals only. The terminal elements represent the dataset attributes. In this paper, the terminal elements represent the genes in microarray datasets. We used the term feature to represent the gene in microarray dataset to avoid the possible confusion between the gene in microarray datasets and the gene in GEP chromosome. The size of the chromosome (ch) is as follows:

$$ch = gene\ length * number\ of\ genes$$
$$gene\ length = h + t \qquad\qquad (2)$$
$$t = h(n - 1) + 1$$

where n represents the maximum number of parameters needed in the function set. The number of genes and h should be given in the initialization process.

The second part of GEP is a genotype which is inherited from GP. The genotype, which is also called expression tree (ET), is created by converted the phenotype (chromosome) into a tree structure using a special language invented by the GEP author.

GEP process includes several steps. It starts with initializing the GEP parameters such as the function set, the terminal set, the number of genes in each chromosome, the head size… etc. The second step is initializing the population, and then in step three the individuals (chromosomes) are evaluated using the suitable fitness function. All the exiting fitness functions defined by GEP are not suitable for gene selection problem, so in this paper we propose a new fitness function (see Sect. 2.4). In step four, the fitness function result is evaluated based on the GEP conditions. The condition can be based on the number of the generations or the evaluation results. If the condition is satisfied the GEP will stop, otherwise GEP process will continue with step five which applies the genetic operations to modify the individuals. The modification is made by the methods of mutation, transposition, root transposition, gene transposition, gene recombination, as well as recombination of one- and two-points. The last step is to prepare the next generation from the modified individuals. The flow chart of the GEP process steps is shown in Fig. 1.

We chose GEP model to create new feature selection method for two main reasons:

- Effectiveness: GEP has the ability to find the solution from a complex environment in a simple way by using the genotype/phenotype representation.
- Flexibility: Each step in GEP process can be changed to suit the adopted problem and environment without adding any complexity to the overall process.

Fig. 1 The flowchart of the
GEP process steps

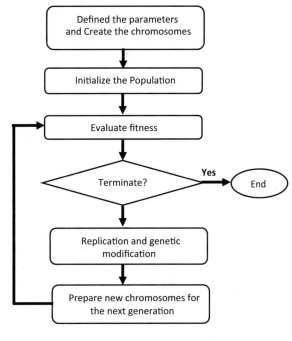

Fig. 2 Stricture of SVM
[32]

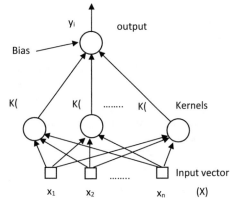

2.3 SVM for Classification

The Support Vector Machine (SVM) is an advanced method of classification [31].
The basis for SVM theory is based on the structural risk minimization (SRM) idea.
The architecture of SVM is shown in Fig. 2.

In Fig. 2, The ith vector in a dataset $\{(xi, yi)\}$, i = 1 to n is symbolised by the
notation xi. yi is the label associated with xi. xi is also known as patterns, inputs, as

well as examples. The kernel functions of the real data are represented by K (Y is the output class label where yi ∈ Y).

The support vector machine (SVM), upon making a training set of instance-label pairs and Y available, needs the solutions of following optimisation problems. The mapping of the training vectors into a higher dimensional space is made by the function.

Binary classification is the simplest form of predicting problems, in an attempt to distinguish between objects belonging to any of the two categories of positive (+1) or negative (−1). Two techniques—large-margin separation and kernel functions—are utilised by SVMs in solving this problem.

The input is first mapped by SVM into a high-dimensional feature space, after which, SVM searches for a separating hyper plane that has the margin maximised between two classes within this space. The maximisation of the margin happens to be a quadratic programming (QP) problem and is possible to be solved from its dual problem through the introduction of Lagrangian multipliers. The SVM, in the absence of any mapping knowledge, utilises the dot product functions known as kernels in feature space to discover the optimal hyper plane. It is possible to write the optimal hyper plane solution of a few input points known as support vectors. SVMs are part of the general kernel methods category. A kernel method can be defined as an algorithm that is dependent on the data by the way of dot-products alone. It is possible for a kernel function to undertake dot product computation in some probable high dimensional feature space when this becomes the case.

There are four basic kernels: Linear, Polynomial, Quadratic and Radial Basis Function (RBF). Selecting the correct kernel function is essential, considering the fact that the feature space where the classification of the training set examples will be made is defined by the kernel function.

In this paper, we used SVM multi classification for several reasons:

- Support Vector Machine is widely used to solve the problem of multi-classification due to its high precision and ability to handle high dimensional data such as gene expressions, as well as flexibility in diverse data sources modelling [23, 24].
- The original GEP is not designed to solve the problem of multi-classification. As a result, we replaced the binary classification in GEP with SVM classification.

The parameters of the SVM classifier in our experiment were set as follows:

- Type of SVM model: C-SVC.
- Kernel function: radial basis function (RBF).
- C search range: 0.1–5000
- Gamma search range: 0.001–50.
- Stopping criteria: 0.001000.
- Cache size: 256.0.

2.4 Hybrid GR-IGEP

In this study, the Gain Ratio (GR) and Improved Gene Expression Programming (IGEP) methods are combined to select relevant genes/features from microarray datasets for classification. In the first-stage, GR, a filter method, is used to reduce the noise by reducing the number of the irrelevant genes. Initially, we calculated the Gain Ratio values (GR values) for eight gene expression data sets by Weka [28]. GR values were calculated for all genes in the microarray data sets and then the feature genes were sorted according to their GR values. A feature with a higher GR value indicates higher discrimination of this feature compared to other categories and means that the feature contains useful information for classification.

After calculating the GR values for all features, a threshold for the results is to be established. Since our results showed that most GR values were zero after the computation process, many features have no influence on the category in a data set, indicating that these features are irrelevant for classification. The threshold in our study is zero for the all data sets. If the information gain value of a feature is higher than the threshold, the feature is selected.

In this paper, IGEP is developed to find the best subset of features from microarray datasets. To do so, we need to take the following major steps. The first step is to define a set of functions and a set of terminals. From our experiments, we find that using many functions does not enhance the algorithm performance but conversely, it may lead to a very complex model with low performance results. For this reason, our function set contains a few arithmetic operators, which are $+$, $-$, $\times$, $\div$, sqr. The terminal set is the selected genes from the microarray data set via GR method above. The second step is to set the initial population, which is formed by randomly creating a group of chromosomes. The parameters used in our IGEP model are listed in Table 1.

In the third step, the generated individuals/chromosomes of the population are evaluated in terms of the fitness. The fitness function is predefined based on the problems concerned. Those chromosomes with a higher fitness are selected as the eugenic ones to generate a new generation via predefined genetic operations. In this paper, we define a new fitness function as follow that is suitable to the feature selection problem.

$$f_i = 2r * AC(i) + r * \frac{f - s_i}{f} \qquad (3)$$

This function consists of two parts. The first part is based on the accuracy result $AC(i)$. This accuracy is provided by support vector machine (SVM) classifiers using LOOCV. The second part is based on the number of the selected features s_i from the total number of features f in the dataset. Parameter r is a random value within the range (0.1, 1) giving an importance to the accuracy with respect to the number of features. Since the accuracy value is more important than the number of selected features, we multiply the accuracy by 2r.

Table 1 Parameters used in the proposed IGEP

Parameter	Setting
Function set	$+, -, \times, \div$, sqr
Terminal set	Selected GR from the microarray dataset
Number of genes in each chromosome	5
Head size	8
Tail size	9
Gene size	17
Linking functions	Addition
Number of chromosomes	200
Maximum Number of generations	2000
Genetic operators	
Mutation	0.044
Inversion	0.1
Gene transposition	0.1
IS transposition	0.1
RIS transposition	0.1
Gene recombination	0.1
One-point recombination	0.3
Two-point recombination	0.3

The predictive accuracy evaluated by the LOOCV method is used to measure the fitness of an individual. In the LOOCV method, a single observation (sample) from the original data is selected as the testing data, and the remaining observations as the training data. This is repeated so that each observation in the dataset is used once as the testing data. Standard genetic operators are applied without modification.

The process of IGEP is repeated until the predefined termination condition is satisfied. A termination conditions in this paper are the number of generations and an expected fitness value. Then the best feature subset is selected from the chromosome which has the best fitness. The flow chart of the GEP process steps is shown in Fig. 3.

3   Experimental Results

The data sets in this study consisted of eight gene expression profiles, which were downloaded from http://www.gems-system.org. The microarray data was obtained by the oligonucleotide technique. The data format is shown in Table 2, which includes the data set name, the number of samples, number of genes (features), Classes, and Reference.

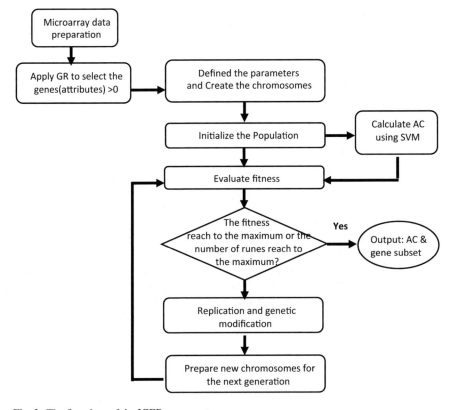

Fig. 3 The flowchart of the IGEP process steps

Table 2 Description of the experimental datasets

Dataset name	Samples	Features	Classes	References
11_Tumors	174	12,533	11	[33]
9_Tumors	60	5726	9	[34]
Brain_Tumor1	90	5920	5	[35]
Brain_Tumor2	50	10,367	4	[36]
Leukemia 1	72	5327	3	[37]
Leukemia 2	72	11,225	3	[38]
Lung_Cancer	203	12,600	5	[39]
Prostate_Tumor	102	10,509	2	[40]

In order to evaluate the performance of our algorithm objectively, we first evaluated the performance in terms of three evaluating criteria; classification accuracy (AC), number of selected attributes (N) and CPU Time (T). Then we compared the

Table 3 Comparison of IGEP with two gene selection algorithms on eight selected datasets

Statistics	PSO	GA	ABC	IGEP
AC avg.	93.295	93.63	93.85125	97.05375
AC std.	0.96625	1.12875	1.19875	0.93
T avg.	130.692463	125.968125	124.74505	247.412025
T std.	4.23975	3309.31515	4.555375	0.033748
N max.	23	546	105	18
N min.	18	533	87	14
N avg.	19	568	176	15
N std.	1.75625	5.1425	5.13125	1.32625

results with two popular gene selection algorithms named Particle Swarm Optimization (PSO) [41], GA [42] and Artificial Bee Colony (ABC) algorithm [43] which uses the same model for the sake of a fair comparison.

The GR algorithm was used in order to filter irrelevant and noisy genes and reduce the computational load for the gene selection and classification methods. The support vector machine (SVM) with a linear kernel served as a classifier of the two gene selection methods. In order to avoid selection bias, the LOOCV was used. Weka software was used to implement the PSO and GA models with default settings, while the IGEP model was implemented by using java package GEP4 J [44]. Table 3 shows the comparison results of IGEP with two gene selection algorithms cross eight selected datasets.

The experimental results showed that the IGEP algorithm achieved the highest average accuracy (ACav) result (97.05375%) across the eight experimental datasets, while the ACavg of other models were 93.295, 93.63 and 93.85125 for PSO, GA and ABC respectively.

The average of the standard deviation (ACstd.) results showed that IGEP had the smallest value (0.93), while the ACstd were 0.96625, 1.12875 and 1.19875 for PSO, GA and ABC respectively. This means the IGEP algorithm made the classification performance more accurate and stable.

The IGEP algorithm achieved the smallest average number of predictive genes (Navg.) was 15, while the Navg. were 18, 533 and for PSO, GA and ABC respectively. This shows that IGEP is a promising approach for gene selection with the smallest number of selected genes.

CPU Time results showed that IGEP time (T) is longer compared with the other methods (PSO, GA and ABC). Table 4 presents the performance evaluation values in terms of AC, N and T for all experimental models on the eight experiment datasets.

Table 4 Compression of the GEP performance with representative gene selection algorithms

11_Tumors	IG-PSO	IG-GA	ABC	IG-GEP
AC avg.	93.06	92.53	94.06	93.88
AC std.	3.01	3.11	3.2	3
T avg.	122.7431	104.756	121.3591	285.1422
T std.	1.2484	2.4632	1.321	0.001264
N max.	42	485	51	26
N min.	31	477	42	11
N avg.	37.5	479	46.8	18.6
N std.	2.8	4.6	3.36	3
9_Tumors	*IG-PSO*	*IG-GA*	*ABC*	*IG-GEP*
AC avg.	75.5	85	87.66	89.83
AC std.	1.3	1.9	1.29	1.01
T avg	173.863	174.573	170	280.114
T std	6.4256	26,443	7.321	0.00221
N max.	31	60	98	23
N min.	26	48	57	18
N avg.	28.8	52	14.9	20.3
N std.	2.3	5.6	1.13	2.1
Brain_Tumor1	*IG-PSO*	*IG-GA*	*ABC*	*IG-GEP*
AC avg.	94.11	93.33	90.11	96.11
AC std.	1.4	1.52	1.6	1.41
T avg.	145.8793	137.931	122.7013	256.346
T std.	4.235	5.642	4.385	0.01641
N max.	22	251	30	20
N min.	18	238	21	18
N avg.	19.1	244	25.5	19
N std.	1.79	5.46	3.47	1.05
Brain_Tumor2	*IG-PSO*	*IG-GA*	*ABC*	*IG-GEP*
AC avg.	89.8	88	88.8	99.8
AC std.	1.01	1.05	1.05	1.01
T avg.	119	121.467	121.467	215.315
T std.	3.525	5.362	5.362	0.0286
N max.	16	490	60	15
N min.	13	488	54	13
N avg.	14.7	489	57.1	14.6
N std.	1.34	0.81	2.42	0.7

(continued)

Table 4 (continued)

Lung_Cancer	IG-PSO	IG-GA	ABC	IG-GEP
AC avg.	98.56	95.57	95.57	98.48
AC std.	0.6	0.1	0.1	0.61
T avg.	130.947	121.972	121.821	235.195
T std.	3.525	5.362	5.362	0.0286
N max.	17	2104	74	15
N min.	12	2099	71	13
N avg.	14.3	2101	71.7	14.5
N std.	2.37	1.41	1.51	0.61
Leukemia1	IG-PSO	IG-GA	ABC	IG-GEP
AC avg.	100	100	100	100
AC std.	0	0	0	0
T avg.	121.75	123.467	120.279	243.535
T std.	5.362	3.525	3.525	0.096
N max.	10	79	75	8
N min.	7	40	57	6
N avg.	8.1	56	87.9	7.7
N std.	0.88	15.42	13.53	0.67
Leukemia2	IG-PSO	IG-GA	ABC	IG-GEP
AC avg.	99	98.61	98.61	100
AC std.	0	0.3	1.3	0
T avg.	89.957	90.648	89.402	203.535
T std.	5.362	3.525	3.525	0.096
N max.	10	785	785	9
N min.	6	780	780	4
N avg.	7.5	782	782	7.5
N std.	1.52	1.83	1.83	1.58
Prostate	IG-PSO	IG-GA	ABC	IG-GEP
AC avg.	96.33	96	96	98.33
AC std.	0.41	1.05	1.05	0.4
T avg.	141.8793	132.931	130.931	260.114
T std.	4.235	5.642	5.642	0.0009
N max.	20	355	350	19
N min.	17	339	304	17
N avg.	18.3	343	324	18.1
N std.	1.05	6.01	13.8	0.9

4 Conclusion

In this paper, an innovative gene selection model GR-GEP is proposed for selecting informative and relevant genes from microarray data sets. The experimental results show that the proposed GR-GEP can provide a smaller set of reliable genes and achieve higher classification accuracies. Furthermore, the comparisons with two gene selection methods on eight considered gene expression datasets show that our proposed GR-GEP can achieve seven out of eight best results in terms of classification accuracy and the number of selected genes. However, GR-GEP spend longer time than other models to reach to the best results. Therefore, how to reduce the processing time for GR-GEP to get higher efficiency and effectiveness is one of our research directions in the near future.

References

1. Chuang, L.-Y., Ke, C.-H., Yang, C.-H.: A hybrid both filter and wrapper feature selection method for microarray classification. arXiv:1612.08669 (2016)
2. Guo, S., et al.: A centroid-based gene selection method for microarray data classification. J. Theor. Biol. **400**, 32–41 (2016)
3. Dashtban, M., Balafar, M.: Gene selection for microarray cancer classification using a new evolutionary method employing artificial intelligence concepts. Genomics **109**(2), 91–107 (2017)
4. Yang, C.-H., Chuang, L.-Y., Yang, C.H.: IG-GA: a hybrid filter/wrapper method for feature selection of microarray data. J. Med. Biol. Eng. **30**(1), 23–28 (2010)
5. Chinnaswamy, A., Srinivasan, R.: Hybrid feature selection using correlation coefficient and particle swarm optimization on microarray gene expression data. In: Innovations in Bio-Inspired Computing and Applications, pp. 229–239. Springer (2016)
6. Algamal, Z.: An efficient gene selection method for high-dimensional microarray data based on sparse logistic regression. Electron. J. Appl. Stat. Anal. **10**(1), 242–256 (2017)
7. Lu, H., et al.: A hybrid feature selection algorithm for gene expression data classification. Neurocomputing (2017)
8. Pino Angulo, A.: Gene selection for microarray cancer data classification by a novel rule-based algorithm. Information **9**(1), 6 (2018)
9. Jain, I., Jain, V.K., Jain, R.: Correlation feature selection based improved-binary particle swarm optimization for gene selection and cancer classification. Appl. Soft Comput. **62**, 203–215 (2018)
10. Cheng, Q., Zhou, H., Cheng, J.: The fisher-markov selector: fast selecting maximally separable feature subset for multiclass classification with applications to high-dimensional data. IEEE Trans. Pattern Anal. Mach. Intell. **33**(6), 1217–1233 (2011)
11. Chuang, L.-Y., Yang, C.-H., Yang, C.-H.: Tabu search and binary particle swarm optimization for feature selection using microarray data. J. Comput. Biol. **16**(12), 1689–1703 (2009)
12. Ferreira, C.: Gene expression programming in problem solving. In: Soft Computing and Industry, pp. 635–653. Springer (2002)
13. Azzawi, H., Hou, J., Xiang, Y., Alanni, R.: Lung cancer prediction from microarray data by gene expression programming. IET Syst. Biol. (2016)
14. Yu, Z., Lu, H., Si, H., Liu, S., Li, X.: A highly efficient gene expression programming (GEP) model for auxiliary diagnosis of small cell lung cancer. PLoS ONE **10**(5), e0125517 (2015)
15. Peng, Y.Z., Yuan, C.A., Qin, X., Huang, J.T., Shi, Y.B.: An improved Gene Expression Programming approach for symbolic regression problems. Neurocomputing **137**, 293–301 (2014)

16. Kusy, M., Obrzut, B., Kluska, J.: Application of gene expression programming and neural networks to predict adverse events of radical hysterectomy in cervical cancer patients. Med. Biol. Eng. Comput. **51**(12), 1357–1365 (2013)

17. Yu, Z., Chen, X.Z., Cui, Si, H.Z.: Prediction of lung cancer based on serum biomarkers by gene expression programming methods. Asian Pac. J. Cancer Prev. **15**(21), 9367–9373 (2014)

18. Alanni, R., Hou, J., Abdu-aljabar, R., Xiang, X.: Prediction of NSCLC recurrence from microarray data with GEP. IET Syst. Biol. **11**(3), 77–85 (2017)

19. Azzawi, H., Hou, J., Alanni, R., Xiang, Y.: Multiclass lung cancer diagnosis by gene expression programming and microarray datasets. In: International Conference on Advanced Data Mining and Applications. Springer (2017)

20. Tan, P.L., Tan, S.C., Lim, C.P., Khor, S.E.: A modified two-stage SVM-RFE model for cancer classification using microarray data. In: International Conference on Neural Information Processing. Springer (2011)

21. Martínez, J., Iglesias, C., Matías, J.M., Taboada, J.M., Araújo, M.: Solving the slate tile classification problem using a DAGSVM multiclassification algorithm based on SVM binary classifiers with a one-versus-all approach. Appl. Math. Comput. **230**, 464–472 (2014)

22. Afshar, H.L., Ahmadi, M., Roudbari, M., Sadoughi F.: Prediction of breast cancer survival through knowledge discovery in databases. Glob. J. Health Sci. **7**(4), 392 (2015)

23. Le Thi, H.A., Nguyen, M.C.: DCA based algorithms for feature selection in multi-class support vector machine. Ann. Oper. Res. **249**(1), 273–300 (2017)

24. Rajaguru, H., Ganesan, K., Bojan, V.K.: Earlier detection of cancer regions from MR image features and SVM classifiers. Int. J. Imaging Syst. Technol. **26**(3), 196–208 (2016)

25. Priyadarsini, R.P., Valarmathi, M., Sivakumari, S.: Gain ratio based feature selection method for privacy preservation. ICTACT J. Soft Comput. **1**(04), 20011 (2011)

26. Karegowda, A.G., Manjunath, A., Jayaram, M.: Comparative study of attribute selection using gain ratio and correlation based feature selection. Int. J. Inf. Technol. Knowl. Manage. **2**(2), 271–277 (2010)

27. Yang, P., Zhou, B., Zhang, Z.: A multi-filter enhanced genetic ensemble system for gene selection and sample classification of microarray data. BMC Bioinform. **11**(1), 1 (2010)

28. Witten, I.H., et al.: Data Mining: Practical Machine Learning Tools and Techniques. Morgan Kaufmann (2016)

29. Golberg, D.E.: Genetic Algorithms in Search, Optimization, and Machine Learning, p. 102. Addison Wesley (1989)

30. Koza, J.R.: Genetic Programming: On the Programming of Computers by Means of Natural Selection, vol. 1. MIT Press (1992)

31. Hearst, M.A., et al.: Support vector machines. IEEE Intell. Syst. Appl. **13**(4), 18–28 (1998)

32. Vanitha, C.D.A., Devaraj, D., Venkatesulu, M.: Gene expression data classification using support vector machine and mutual information-based gene selection. Procedia Comput. Sci. **47**, 13–21 (2015)

33. Su, A.I., Welsh, J.B., Sapinoso, L.M.: Molecular classification of human carcinomas by use of gene expression signatures. Can. Res. **61**(20), 7388–7393 (2001)

34. Staunton, J.E., et al.: Chemosensitivity prediction by transcriptional profiling. Proc. Natl. Acad. Sci. **98**(19), 10787–10792 (2001)

35. Pomeroy, S.L., et al.: Prediction of central nervous system embryonal tumour outcome based on gene expression. Nature **415**(6870), 436 (2002)

36. Nutt, C.L., et al.: Gene expression-based classification of malignant gliomas correlates better with survival than histological classification. Can. Res. **63**(7), 1602–1607 (2003)

37. Golub, T.R., Slonim, D.K., Tamayo, P.: Molecular classification of cancer: class discovery and class prediction by gene expression monitoring. Science **286**(5439), 531–537 (1999)

38. Armstrong, S.A., et al.: MLL translocations specify a distinct gene expression profile that distinguishes a unique leukemia. Nat. Genet. **30**(1), 41 (2002)

39. Bhattacharjee, A., Richards, W.G., Staunton, J.: Classification of human lung carcinomas by mRNA expression profiling reveals distinct adenocarcinoma subclasses. Proc. Natl. Acad. Sci. **98**(24), 13790–13795 (2001)

40. Singh, D., et al.: Gene expression correlates of clinical prostate cancer behavior. Cancer Cell **1**(2), 203–209 (2002)
41. Moraglio, A., Di Chio, C., Poli, R.: Geometric particle swarm optimisation. In: European Conference on Genetic Programming. Springer (2007)
42. Goldberg, D.E.: Genetic Algorithms in Search, Optimization and Machine Learning. Addison-Wesley, Reading, MA (1989)
43. Karaboga, D., Basturk, B.: Artificial bee colony (ABC) optimization algorithm for solving constrained optimization problems. In: International Fuzzy Systems Association World Congress. Springer (2007)
44. Thomas, J.: GEP4J (2010)

A Novel Differential Selection Method Based on Singular Value Decomposition Entropy for Solving Real-World Problems

Rashmi and Udayan Ghose

Abstract An evolutionary singular value decomposition (SVD) entropy based feature selection approach is proposed for finding optimal features among large data sets. Since the data typically consists of a large number of features, all of them are not optimal. In this paper, an optimal feature selection approach based on differential evolution (DE) and SVD entropy is proposed. The functioning of the proposed approach is examined on available UCI data sets. This approach provides ranked features by optimizing SVD entropy using the DE. An SVD entropy based fitness function is employed as the criterion to measure the optimal features and this makes the new approach easier to implement. DE results in a faster and accurate convergence towards global optima. The proposed approach shows its effectiveness on binary data sets with a number of features ranging between 9 and 60. The result explains that the proposed approach can converge quickly and rank the features. The experimental section demonstrates the results in terms of classification accuracy by Support Vector Machine (SVM) and Naive Bayes (NB) classifiers. The explored results are favorable and strengthen the contribution of the proposed approach.

1 Introduction

Finding optimal features to improve efficiency, scalability and accuracy of classification process is essential [1, 2]. To find optimal feature is a difficult task. In past years, this difficulty has motivated extensive research effort focusing on speeding the search process. This has been resolved to a certain limit by resting the rigidity of an

Rashmi (✉)
University School of Information Communication and Technology,
Amity School of Engineering and Technology, Guru Gobind Singh Indraprastha University,
Dwarka, New Delhi, India
e-mail: rashmibehal@gmail.com

U. Ghose
University School of Information Communication and Technology, Guru Gobind Singh
Indraprastha University, Dwarka, New Delhi, India
e-mail: udayan@ipu.ac.in

© Springer Nature Switzerland AG 2019
R. Lee (ed.), *Computer and Information Science*, Studies in Computational
Intelligence 791, https://doi.org/10.1007/978-3-319-98693-7_3

optimal solution to gratify the feature selection problem [3, 4]. Heuristic measures help to identify the search space which explores the optimal solution [4]. Throughout the search methods that compete to recede the computational burden with optimism, the DE and its ancestors have been receiving the most attention. While the DE, SVD, and SVD entropy are well-known terms and considered one of the essential tools in data reduction.

Storn and Price [5] proposed the DE scheme, it is comparatively simple, fast and population based stochastic search technique. DE plunge under the category of Evolutionary Algorithms (EA) with different schemes; and each of them has its own significance. As the trial vector generation process uses the distance information and its direction of the present population to originate a trial vector, thus crossover is practiced to originate a trial vector, which is further used in the period of mutation operation to originate one successor while, in DE, cross over is followed with mutation. The application of the heuristic search allows the user to retrieve the optimal features representing the feature set optimization process. In tune with SVD, this reduces the number of features and DE have to explore the optimal solution using the least number of features and improves classification accuracy. Substantial effort have been infused into the fastening of the DE algorithm over the years. Many updated versions of algorithms have been defined, gaining speed or other benefits have been defined usually by relaxing the DE optimal concept.

SVD is a linear map of the class data from the complete data, the vector space is used to reduced eigenfeatures and eigenarrays space. In this stretch, the data is diagonalize in such a way that each eigen-feature is expressed only in the corresponding eigen-array, with the analogous eigen expression level which implies their concerned significance. This data driven statistics is inherited from the eigenfeatures and eigenarrays due to the orthogonal superposition of the features and arrays. This SVD value helps in computing SVD entropy. SVD Entropy measures feature relevance using an entropy measure based on the Singular Value Decomposition (SVD) of the data matrix [6]. Alter et al. define the SVD based entropy of the data set. This entropy also varies between 0 and 1. This SVD entropy value corresponds to order/disorder of a data set which can is explained by a single eigenvector. The low entropy value shows about non uniform behaviour of the eigenvalues. This entropy is different from the Shannon entropy as it is based on probabilities while SVD entropy is based on the distribution of eigenvalues. The contribution of ith feature to the entropy by max fitness function according to DE. Thus, ranked features are retrieved by the SVD entropy fitness function. Concurrently, the optimality feature selection process was put into hybrid concept of fuzzy entropy [7].

This work aims to fasten the DE by introducing the notion of SVD entropy, which would result in substantial computational time. Thus, in this paper, SVD entropy is used in analyzing the fitness of the data. In this paper, the concept of SVD entropy with DE is proposed. The main objective is to minimize redundancy, uncertainty and improve accuracy of the data set. Section 2 describes about the basic concepts of differential evolution and SVD entropy is used in finding optimal features. Section 3 outlines the proposed approach. Section 4 informs about the data sets used and experimental results achieved. Section 5 concludes the work.

2 Preliminaries

2.1 Differential Evolution (DE)

DE is an effective evolutionary optimization technique. In [5] Storn, proposed a population based global optimization algorithm. The initial population is created by NP candidate solutions, known as individuals. The initial population covers almost complete search space. The variety and selection process drives evolution. The variance process explore individual fields of the stretch space while previous experiences help in the exploitation of the selection process. It is easy to use, good convergence, and fast implementation properties [8]. Its full process is explained in the subsections.

Table 1 shows a dataset as $IS = (U, A, V, f)$, where $U = \{x_1, x_2, ..., x_n\}$, is a non empty finite set of objects, called as universe of discourse, $A = \{a_1, a_2, ..., a_n\}$, is a non empty finite set of attributes, called as features, V is union of feature domain $V = \cup a \in A \ V_a$, for V_a denoting value domain for feature a $f : U \times A \rightarrow V$.

2.1.1 Mutation Operation

The generations in DE is used to find the new population by $gen_0, ..., gen_{max}$. Generally each individual act as a D-dimension vector named as a target vector $T_{i,gen} = T_{i,gen1}, ..., T_{i,genn}$. At new generation (time) t the ith individual of the population with the optimization factors CR and F. Every generation updates the population, hence a better donor vector is created. The donor vector creation scheme distinguishes the existing DE scheme. Here DE earliest variant DE/rand/bin/1 scheme is used to compute a mutation operation to produce a mutant operator $M_{i,gen}$ corresponding to each individual $T_{i,gen}$ in the present population. Target vector $T_{i,gen}$ at the generation gen, and its mutant vector $M_{i,gen} = M_{i,gen1}, ..., M_{i,genn}$ is produced for each ith member, other three parameter vectors r_1, r_2, r_3 [1, NP] with inequality among all are picked up randomly from the present population. The donor vector

Table 1 Information system (IS)

Sr. No	Conditional attributes					Decision attribute
	a	b	c	d	e	D
x1	v11	v12	v13	v14	v15	d1
x2	v21	v22	v23	v24	v25	d2
x3	v31	v32	v33	v34	v35	d2
x4	v41	v42	v43	v44	v45	d1
x5	v51	v52	v53	v54	v55	d2

$M_i(t)$ is formulated by a scalar value F with the difference of any two parameter vectors. The new vector for k may be expressed as

$$M_1(t) = Tr_{1,k}(t) + F(Tr_{2,k}(t) - Tr_{3,j}(t))$$ (1)

2.1.2 Binary Crossover Operator

A binary crossover operation followed by the mutation phase, is enforced to improve the potential diversity of the population. Three different crossover schemes, popular as Arithmetic, Binomial, and Exponential can be used. Here, Binomial operator is practiced on an exclusive pair of target vector $T_{i,gen}$ and its corresponding mutant vector $M_{i,gen}$ to produce a new trial vector that is denoted as TR. The best value is kept in target vector T_{gen}. This scheme is performed on each feature with the Crossover Rate (CR) between 0 and 1. CR is a user defined constant which controls the mutation. This scheme is created in following fashion:

$$TR_{i,gen} = \begin{cases} T_{i,gen} & \text{if } random(0, 1) \le CR \\ scheme_{i,gen} & \text{otherwise} \end{cases}$$ (2)

$random(0,1) \, \varepsilon \, [0.1]$ is a uniformly distributed random number, and i is a index, which confirms that $TR_{i,gen}$ takes at least one of the features from $scheme_{i,gen}$.

2.1.3 Selection Operator

This is performed to determine updated trial vector (TR) between TR and T. The selection is done with the help of the fitness values. The selection operator is defined as follows:

$$TR_{i,gen+1} = \begin{cases} T_{i,gen} & \text{if } f(TR_{i,gen}) \le f(T_{i,gen}) \\ TR_{i,gen} & \text{otherwise} \end{cases}$$ (3)

where f(x) is the objective function computed using decision variable X and i = 1, 2, …, Np. This is a one to one selection like greedy approach to select the global optimum.

2.2 Singular Value Decomposition Entropy

SVD is well popular term in linear algebra. Golub et al. [9] proposed a low rank approximation technique known as SVD, which helps in data reduction. Necessarily, SVD factors a matrix M into a product of three matrices U, D, V. The SVD of the data matrix is defined as follows in Eq. (4):

$$M = UDV^t \tag{4}$$

The small value of U shows the better value. Hence, each eigenvalue of the matrix M be denoted by M_j^2 or eigen array is spotted from the other eigenvalues and eigenarrays. The normalized eigenvalues are defined as in Eq. (5),

$$V_j = M_j^2 / \sum_j M_j^2 \tag{5}$$

Entropy is used to measure uncertainty of information in the system. It can also be measured by rough entropy [10], and combination of fuzzy and Shannon entropy [11] Data set randomness can be measured using Shannon entropy. Shannon entropy measures information of each attribute by the formula given in [12] shown below Eq. (6)

$$H(x) = -\sum_{i=1}^{N} P_i \log P_i \tag{6}$$

In the existing literature SVD entropy of the data set was defined by [6, 13] shown in Eq. (7)

$$E(x) = -\frac{1}{\log(N)} \sum_{j=1}^{N} V_j \log V_j \tag{7}$$

where N is the rank of the data matrix. This is used to normalized the data.

2.3 Support Vector Machine

Support vector machine (SVM) is a supervised machine learning algorithm developed by [14]. It is used for classification problems. It is usually used for linear classification but it can also perform non linear classification using kernel trick. It creates a hyper plane or a set of hyper plane in a high dimensional space which can be used for classification. These hyper planes are used to separate the two classes. This support vector algorithm is used for an optimal separating hyper plane that maximizes the separating margin between the two classes of data. Since the larger margin can acquire, the better generalization ability.

2.4 Naive Bayes Classifier

Naive Bayesian classifier is based on the information present in the data. All attributes are independent of each other. It is simple, fast, accurate and reliable. It is based on simplifying assumption that the attribute value is conditionally independent with the given target value. It is a probabilistic algorithm to classify the data. Probabilistic means that they calculate the probability of each class for the given data. These probabilities are obtained by using Bayes theorem for each feature on the basis of prior knowledge of the conditions that might be related to that feature. Formulating prior probability, helps in classifying. The classifier has the number of probabilities that must be estimated from the training data attribute values. It must be a smallest number rather than to estimate all $P(a_1, \ldots, a_n \mid v_j)$. The set of these probabilities corresponds to learned interpretation. This interpretation is used to classify each new instance by using the equation:

$$value_{NB} = argmax_{v_j \varepsilon value} P(value_j) \prod_i P(a_i \mid value_j) \qquad (8)$$

This classifier has no explicit search through the space of possible interpretations. Probability is used in interpretations rather than any search in data [15].

3 Proposed Method

The proposed method follows the following steps: (1) Preprocessing is performed using the standard deviation of data, (2) Generating population using differential evolution, and (3) finding fitness of data using SVD entropy and ranking the features, (4) Post processing using SVM and NB classifier based on optimal features. Figure 1 shows the pseudo code for proposed optimal feature selection method. The SVD entropy pseudo code is depicted in Fig. 2.

INPUT: Original Data Set(ODS)
OUTPUT: Optimal Features(OF)

1. **Selection** Initially, select the data of a particular class.
2. **Normalization** Normalize the data with standard deviation
3. **Generation** Generate random population with minimum and maximum value in data.
4. **Fitness Evalution** Fitness is computed using function SVD_Entropy
5. **Population** New population is generated with fitness of parent and child
   if SVD_Parent < SVD_Child

SVD_Parent = SVD_Child
New_Population = Population_Parent

6. **Ranking Feature** Feature with maximum SVD_Parent value is the optimal one.

Fig. 1 Proposed method

INPUT: Normalize Data Matrix
OUTPUT: Entropy Vector

1. **Computation** Compute SVD of the matrix using Eq. (1) - (7)
2. **Find**Find the rank of the matrix
3. **Compute** Entropy is computed as total sum of singular matrix is divided by the sum of the diagonals element.
4. **Return** return the entropy vector.

Fig. 2 SVD_Entropy

3.1 DE-SVDE

In this section, we demonstrate the proposed method incorporated with DE and SVD entropy. This is proposed due to DE convergence towards global optima, SVD eliminates redundancy and SVD entropy measures the meaningful information. Feature ranking using the combination of all these is more optimal. The proposed method removes the redundant and non-optimal features.

In contrast to the basic DE algorithm, the proposed method evaluated fitness from the singular value decomposition entropy of the full feature set. For the feature has to be higher ranked it must have higher entropy, among the non-ranked feature set and it is selected simply by Eq. 7.

SVDE procedure can be used to perform feature ranking. But DE can be performed in a more efficient way on given SVD entropy (see, for example, Eqs. (3) and (7)).

3.2 Monitoring the Search Procedure: Novel Parent Selection

The Proposed method is mainly addressing the problem that how to select parent for mutation of each individual. DE generates random population using formulated equation as stated in Eq. (9) which is based on Eq. (3) to be used as the fitness function in DE to guide the search to find optimal feature subsets. With respect to the target individual $X_{i,gen}$, the selection entropy $E_{i,j,gen}$ of $X_{j,gen}$ is set as its parent value$_i$ ($X_{i,gen}$). In order to ensure that the entropy of all the individuals are normalized, this is retrieved by dividing the by rank of matrix as shown in Eq. (7). The new population selection is done with the help of the fitness function. The selection function is defined as follows:

$$TR_{i,gen+1} = \begin{cases} T_{i,gen} & \text{if } f(E(TR)_{i,gen}) \leq f(E(T)_{i,gen}) \\ TR_{i,gen} & \text{otherwise} \end{cases} \qquad (9)$$

where f(x) is the fitness function computed using SVD entropy as defined in Eq. (7). This function is minimized using the minimum entropy value. Whichever is better in target vector and trial vector is stored in target vector. This process continues over

Fig. 3 Proposed method
flowchart

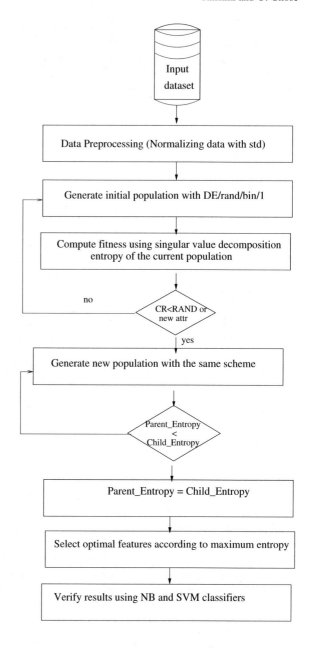

the generations until the best fitness value is obtained. This entropy is used to find the fitness of data. The entropy is computed for each jth feature, $j = 1 \dots D$, of each new generation by differential evolution. Of course the promising individuals with better fitness and near to target individual have more hope to be selected as parent. A novel method named DESVDE is proposed with the use of DE as the search method using Eq. (9) as the fitness function to find optimal feature subsets. The representation of an individual is a M-bit continuous vector representing a possible feature subset where the possible values in the vector is in the range of [0, 1]. Note that the other fitness models can be used in the proposed method. The effects of other fitness models will be studied in the future work (Fig. 3).

3.3 Complexity Analysis

In comparison with the basic DE algorithm, the additional computation of DE-SVDE depends on SVDE. The position measures in each generations are used to evaluate the effectiveness of each individual. For the novel parent selection using SVDE, the selection probabilities computed are based on matrices $O(N^3)$ and the complexity by DE is $O(G \cdot NP \cdot D)$, where G represents the maximum number of generations, D is the dimension and NP is the population size. Hence the proposed method has the total complexity of $O(G \cdot NP \cdot D + 1)$ is same in proportion to N_3 but with better results.

4 Experimental Design and Results

4.1 Parameter Settings

In this work, performance of SVDE based feature ranking algorithm is tested on nine different datasets from the UCI machine learning repository [16]. The data sets are chosen so that they have distinct number of features and instances. Generally, the population size NP is set to be proportional to the dimension D in DE literature [5], D is the number of attributes in the data set. After testing, the parameter values of DESVDE algorithm are set as follows:

1. The initial values of F and CR are set to 0.5 and 0.9 respectively.
2. The NP_{max} are set to 10 * D.
3. The number of generations of DE gen is 30.

4.2   SVDE Results

In this section, the proposed method performance is evaluated in two main perspectives (1) accuracy of selected features and (2) ranking of estimated features in
section (Sect. 1). Details of the chosen synthetic data is shown in section (Sect. 2)
Table 3 shows the optimal features with their ranking. The SVM and NB accuracy are
computed for original and selected ranked feature data. The comparative results are
corroborated in terms of accuracy noticed in the next section. No significant change
was observed in the performance using different seeds. The objective is to search
features with maximum rank, means, according to maximal contribution to underlying overall characteristics. The threshold for maximal rank was set to minimum of
entropy for all data sets.

Nine real data sets, taken from the UCI machine learning repository [16], are
considered. Here, we measured the performance of the method for reliability by
different classifiers SVM and NB. The small number of generations selects a large
number of features and high number of generations selects few dimensions. Clearly,
without sufficient information, resultant feature sets could not assure a respectable
classy performance. Thus, a relative medium number of generations are preferred.
Also, runs should be a moderate value to determine when the selection process should
be stopped. Ideally runs should be large to find the robust fitness selection. In this
paper, the results are shown in Table 3 on nine binary data sets with 30 generations
and ten runs.

DE/Rand/bin/1 scheme and SVD entropy is used to compute the threshold levels
effectively. The DE parameters are set up as guidelines are provided in the literature [5] Results of evolutionary algorithm have been provided as the mean of ten
independent runs where every run was continued till the $D \times 30$ number of fitness
is evaluated. The significance of singular value decomposition entropy approach
is achieved by contrasting results with the original data in terms of classification
accuracy.

4.3   Data Sets

4.3.1   Heart Data Set

Heart data set consists of 14 attributes with 276 instances. This dataset is composed
of a number of tests for healthy people and a heart disease person. In this data age,
blood pressure, maximum heart rate, serum cholesterol, old peak, the number of
major vessels are real-valued attributes, and gender, fasting blood sugar, exercise-
induced angina is binary attributes, while chest pain, type resting electrocardiographs
results, that are nominal attributes.

When the accuracy is tested on the original data set, it is 80.37 and 66.67% while
the classification accuracy of optimal data set is much better which is 80.39 and

74.07%. After selecting the optimal features, the number of features is 7% of the complete data. The results are verified by both the classifiers.

4.3.2 Parkinsons Data Set

The data set is used to check whether the person is healthy or not. Since the complete data features are 22 and selected optimal features are 18. This is listed in Table 3 with their ranks. The mean classification accuracy of optimal data are 99.09 and 79.48% which is an improvement over the original data accuracy that is 93.34 and 61.53%.

4.3.3 CRX Data Set

CRX stands for credit card application data. To assure data confidentiality every attribute value is stored with a meaningless symbol. In this data all kinds of attribute are found like continuous, nominal, where some of the nominal attributes have large value and has some small value. Missing values are also present, which are replaced by the mean of the available data. The data are distributed as 307 (45.5) cases of positive class and 383 (55.5) of negative class. With optimal features, the mean classification accuracy is 85.65 and 64.07% while with the complete data are 76.07 and 60.67%. There is significant improvement in accuracy with only five features (Table 2).

4.3.4 Pima Data Set

Harris [17] has investigated the Pima Indians at the National Institute of Diabetes and Digestive and Kidney Diseases (NIDDK) since three decades. Concerned analysis of the data states that vigorous predictor of diabetes is an unhealthy weight (6th attribute

Table 2 Original data sets description

Data set	Sample cases	Attributes size	Classes
Heart	270	14	2
Sonar	208	61	2
Pima	768	9	2
Parkinson	195	23	2
Wdbc	569	32	2
Ionosphere	351	18	2
Soybean	307	35	2
Vote	435	17	2
Crx	690	16	2

Table 3 Data set with ranked features

Data set	Input features	Number of ranked features	Features rank list
Heart	13	**3**	1, 2, 13
Sonar	60	**45**	4, 22, 23, 59, 11, 16, 41, 50, 15, 17, 28, 42, 48, 49, 51 52, 18, 26, 37, 46, 55, 2, 6, 7, 8, 40, 43, 1, 9, 30 31, 33, 56, 3, 21, 25, 38, 39, 47, 54, 58, 24, 35, 44, 60
Pima	8	**3**	2,4,6
Parkinson	22	**18**	4, 7, 1, 3, 6, 19, 2, 8 20,10,12,13,15,21,9,14,16,17
Wdbc	32	**25**	22, 7, 24, 11, 1, 9, 12, 14, 21 3, 10, 15, 17, 28, 2, 6, 8, 13 18, 20, 29, 16, 27, 20, 31
Ionosphere	34	**8**	14, 11, 31, 33, 15, 3, 19, 29
Soybean	35	**4**	22, 19, 28, 34
Vote	17	**14**	5, 6, 8, 2, 11, 12, 13, 15, 3, 1, 4, 7, 14, 16
Crx	16	**5**	3, 2, 12, 11, 15

in the complete data). NIDDKs experimental results states that, one another predictor is glucose concentration to develop Diabetes. Harris et al. also states that it is a genetic disease. The data contains the personal data of the Pima Indian population. Table 3 shows the lists of optimal attributes of PIDD. Table 4 verifies results with different classifiers, accuracy are depicted as 80.39 and 91.23% which is much better than complete data.

Table 4 Classification results with SVM and Naive Bayes classifiers on Binary Class Data sets

Dataset	Original data mean accuracy (%)		Reduced data mean accuracy (%)	
	SVM	NB	SVM	NB
Sonar	76.92	70.17	**99.17**	**73.17**
Heart	80.37	66.67	**80.39**	**74.07**
Crx	76.07	60.67	**85.65**	**64.07**
Soybean	**97.39**	88.52	90.31	**88.57**
Vote	96.09	89.66	**99.58**	**90.80**
Ionosphere	96.09	89.66	**99.58**	**90.80**
Wdbc	97.71	97.34	**99.80**	92.03
Parkinsons	93.34	61.53	**99.09**	**79.48**
Pima	74.21	76.47	**80.39**	**91.23**

5 Result Discussion

Results for individual classifiers and proposed approach using UCI data sets are summarized in Table 4. The UCI data sets [16] are widely used as a benchmark for evaluating many algorithms. Here Binary class data are chosen to show the effectiveness of the proposed approach. Simulations are done on optimal features data of each class and original data. SVM and NB classifiers are used to measure the accuracy. It has been a common metric measure for assessing the classifier performance for years [18].

In SVM classifier with k = 10 cross validation selected features are trained with half training samples and tested on the half test samples. Finally, the mean classification accuracy is computed. As the training data increases, Naive Bayes performance increases [19]. Accuracy measures discussed in Sect. 2 evaluates the proposed approach accuracy. The effective values are outlined in Table 4. In addition to classification accuracy, it is also demanding to measure optimal number of features while doing selection: arbitrary fitness is achieved with this evolutionary SVD entropy approach. From the Table 4 it is clear that selected feature data helps in improving accuracy. Which means these selected features are more informative; hence can be used to for complete data classification. Table 3 shows the optimal features with their ranks. An optimal features subset for each of nine data sets Heart, Sonar, Pima, Parkinsons, Wdbc, Ionosphere, Soybean, Vote and Crx are selected. The optimal features for the data sets are 3, 45, 3, 18, 25, 8, 4, 14, and 5 respectively. The proposed method observed accuracy is shown in Table 4 which states that it achieves better accuracy over the complete data. This is true for both the classifiers. The classification results show that the proposed approach either gets an above 90% classification or near to complete data accuracy for both the classifiers with reduced dimensions. Meanwhile, it is also noticed that as the number of features increases accuracy drops. This states that more features does not increase the classification accuracy but, on the contrary, due to a small number of training samples the accuracy decreases. The proposed approach continues to improve the accuracy because it is able to extract more information in terms of class separability than the normal fitness function.

The working of the proposed approach, however, is considerably easy and implementation of the approach evolves the new generations with maximum fitness. The results using these nine data sets are compared with the complete data. The classification accuracy is computed using the formula stated as follows:

$$Accuracy = \left(\frac{TP + TN}{TP + FP + FN + TN} \right) \times 100 \qquad (10)$$

In Eq. (10), TP, represents the number of true positives; TN, the number true negatives; FN, the number false negatives; FP, the number false positives as derived from confusion matrix.

For the classification accuracies comparison, box plot visualization is shown in Fig. 4. The box plots for SVM and Naive Bayes accuracy on optimal and complete

Fig. 4 Accuracy of the proposed method of original and reduced data set using SVM and Naive Bayes Classifier

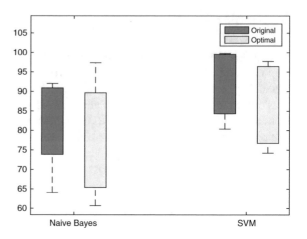

data is shown graphically. Like other visual methods, the box plot, is better representation of a table. This visualization method enhances our understanding of data and help us make comparisons across accuracies [20]. This suggests us accuracy in diverse thoughts about this side or that side. Uneven size sights that diverse data sets have the same contour at some parts of the measure, but in other parts of the scale data sets are more uncertain in their contour. The small upper whisker in the example means that data set contours are stable amongst.

6 Conclusion

This paper proposes an optimal ranked selection method. The proposed method reprocessed the data attribute wise by min, max, and mean and standard deviation mapping function that results in normalized data. These attributes are used to generate an initial population with DE approach. The features fitness is computed by SVD entropy and rank them accordingly. These features may contain the noisy information that is further screened using SVD entropy. The proposed method is analyzed both subjectively and objectively. It is found to be superior as classifiers ac-curacies depicts. The proposed method presented in this paper achieves significant improvement in terms of quality of the data and ranking the features. In the future, the advantage of soft computing along with SVD may be used to develop an optimal feature selection algorithm at the rapid rate with soft computing techniques.

References

1. Han, J., Pei, J., Kamber, M.: Data Mining: Concepts and Techniques. Elsevier (2011)
2. Xing, E.P., Jordan, M.I., Karp, R.M., et al.: Feature selection for high-dimensional genomic microarray data. ICML **1**, 601–608 (2001)
3. Jain, A., Zongker, D.: Feature selection: evaluation, application, and small sample performance. IEEE Trans. Pattern Anal. Mach. Intell. **19**(2), 153–158 (1997)
4. Devijver, P.A., Kittler, J.: Pattern Recognition: A Statistical Approach. Prentice Hall (1982)
5. Storn, R., Price, K.: Differential evolution—a simple and efficient heuristic for global optimization over continuous spaces. J. Glob. Optim. **11**(4), 341–359 (1997)
6. Alter, O., Brown, P.O., Botstein, D.: Singular value decomposition for genome-wide expression data processing and modeling. Proc. Natl. Acad. Sci. **97**(18), 10101–10106 (2000)
7. Chao, Y., Dai, M., Chen, K., Chen, P., Zhang, Z.: Fuzzy entropy based multilevel image thresholding using modified gravitational search algorithm. In: 2016 IEEE International Conference on Industrial Technology (ICIT), pp. 752–757. IEEE (2016)
8. Price, K., Storn, R.M., Lampinen, J.A.: Differential Evolution: A Practical Approach to Global Optimization. Springer Science & Business Media (2006)
9. Golub, G.H., Van Loan, C.F.: Matrix Computations, vol. 3. JHU Press (2012)
10. Rashmi, Ghose, U., Anika: A data reduction method based on indiscernibility and rough entropy for uncertain system. In: International Conference on Intelligent Systems (IntelliSys), pp. 482–487. IEEE (2017)
11. Rashmi, Ghose, U., Mehta, R.: Attribute reduction using the combination of entropy and fuzzy entropy. In: Networking Communication and Data Knowledge Engineering, pp. 169–177. Springer (2018)
12. Shannon, C.E., Weaver, W.: The Mathematical Theory of Communication. University of Illinois Press (2015)
13. Banerjee, M., Pal, N.R., Some modification and extension: Feature selection with SVD entropy. Inf. Sci. **264**, 118–134 (2014)
14. Vapnik, V.: The Nature of Statistical Learning Theory. Springer Science & Business Media (2013)
15. Mitchell, T.M.: Machine Learning. Engineering/Math, vol. 1. McGraw-Hill Science (1997)
16. Newman, D.J., Hettich, S., Blake, C.L., Merz, C.J.: {UCI} Repository of Machine Learning Databases (1998)
17. Demouy, J., Chamberlain, J., Harris, M., Marchand, L.H.: The Pima Indians: pathfinders of health. National Institute of Diabetes Digestive Kidney Diseases, Bethesda, MD Google Scholar (1995)
18. Kodratoff, Y.: Introduction to Machine Learning. Morgan Kaufmann (2014)
19. Rish, I.: An empirical study of the Naive Bayes classifier. In: IJCAI 2001 Workshop on Empirical Methods in Artificial Intelligence, vol. 3, pp. 41–46. IBM (2001)
20. Williamson, D.F., Parker, R.A., Kendrick, J.S.: The box plot: a simple visual method to interpret data. Ann. Internal Med. **110**(11), 916–921 (1989)

Power Consumption Aware Machine Learning Attack for Feed-Forward Arbiter PUF

Yusuke Nozaki and Masaya Yoshikawa

Abstract To prevent semiconductor counterfeits, the physical unclonable functions (PUFs) have attracted attention. Since PUFs utilize the variation of semiconductor manufacturing, physical cloning of PUFs is difficult. However, the risk of machine learning attacks, which clone the function of PUFs, has been reported. In recent years, a new machine learning attack using side-channel information, such as power consumption or electromagnetic wave generated during the operation of the PUF, was reported. Therefore, to consider the security of PUFs in the future, the evaluation of the resistance of PUFs against various attacks is very important. This study proposes a new machine learning attack using power consumption waveforms for the feed-forward arbiter PUF which is one of the typical PUFs. In experiments on a field programmable gate array (FPGA), the validity of the proposed analysis method and the vulnerability of the feed-forward arbiter PUF were clarified.

Keywords Physical unclonable function · Feed-forward arbiter PUF
Machine learning attack · Side-channel attack · Hardware security

1 Introduction

In recent years, the threat of semiconductor counterfeits is pointed out. According to the report [1] of the European Semiconductor Industry Association (ESIA), more than one million semiconductor counterfeits are seized at customs. Since devices embedded semiconductor counterfeits may cause an accident [2, 3], countermeasures are very important. Therefore, physical unclonable functions (PUFs) have been attracted attention as techniques to prevent semiconductor counterfeits [4–8]. The PUF con-

Y. Nozaki (✉) · M. Yoshikawa
Department of Information Engineering, Meijo University, 1-501 Shiogamaguchi,
Tenpaku-ku, Nagoya, Aichi, Japan
e-mail: 143430019@ccalumni.meijo-u.ac.jp

M. Yoshikawa
e-mail: dpa_cpa@yahoo.co.jp

© Springer Nature Switzerland AG 2019
R. Lee (ed.), *Computer and Information Science*, Studies in Computational
Intelligence 791, https://doi.org/10.1007/978-3-319-98693-7_4

verts the variation of semiconductor manufacturing to a unique ID. Several PUFs, including arbiter PUF [4], ring oscillator PUF [5], glitch PUF [6], SRAM PUF [7], pseudo linear feedback shift register PUF [8], and so on, have been proposed. Also, the arbiter PUF is a typical PUF.

Since PUFs utilize the variation of semiconductor manufacturing, physical cloning of PUFs is difficult. However, the risk of machine learning attacks [9–11], which clone the function of PUFs, has been reported. The vulnerability of the arbiter PUF against machine learning attacks is pointed out because it can be represented as a linear model [9]. Therefore, feed-forward arbiter PUF [4], XOR arbiter PUF [4], and lightweight PUF [12], which have resistance against machine learning attacks, have been proposed.

On the other hand, a new machine learning attack using side-channel information, such as power consumption or electromagnetic wave generated during the operation of the PUF, was reported [13, 14]. Therefore, to consider the security of PUFs in the future, the evaluation of the resistance of PUFs against various attacks, including the machine learning attack using side-channel information, is very important.

This study proposes a new power consumption aware machine learning attack for the feed-forward arbiter PUF which is one of the typical PUFs. The proposed analysis method performs the hierarchical machine learning attack using the power consumption generated during the PUF operation. In experiments on a field programmable gate array (FPGA), the validity of the proposed analysis method and the vulnerability of the feed-forward arbiter PUF are verified.

2 Preliminaries

2.1 Physical Unclonable Function

The PUF converts the variation (characteristic value) of semiconductor manufacturing to a unique ID as a digital value. Since the variation of each device is difference, the ID is unique. In the basic operation of the PUF, an input value (challenge) is provided to the PUF, and then an output value (response) is obtained. Also, challenge response pairs (CRPs) are used as the unique ID. Several PUFs, including arbiter PUF [4], ring oscillator PUF [5], glitch PUF [6], SRAM PUF [7], pseudo linear feedback shift register PUF [8], and so on, have been proposed. Then, the arbiter PUF is one of typical PUFs.

2.1.1 Arbiter PUF [4]

The basic structure of the arbiter PUF is depicted in Fig. 1. As shown in Fig. 1, the arbiter PUF consists of two equal length wirings, N selector units, and an arbiter circuit. First, a N-bit challenge as a selection signal is provided to N selector units,

Fig. 1 Structure of arbiter
PUF

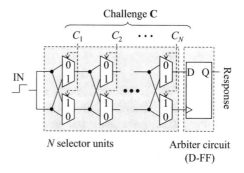

and then two signal propagation paths from an IN signal to the arbiter circuit are determined. At this time, when the challenge is zero, the signal path goes to straight, and when the challenge is one, the signal path crosses. Then, the determined two signals propagate, and they are inputted to the arbiter circuit. In the arbiter circuit, an arrival order of two signals is judged. Here, in the case of using a delay flip-flop (DFF) as the arbiter circuit, two signals are provided to D and CLK signals. Then, when the signal arrives first in D, the response (output of DFF) is judged as one. By contrast, when the signal arrives early in CLK, the response is determined as zero. In the arbiter PUF, since the signal propagation delays differ by the variation of semiconductor manufacturing, the generated response is a unique in each device.

2.1.2 Machine Learning Attack for Arbiter PUF

The machine learning attack models the structure of the arbiter PUF as a linear model [9, 10]. The modeling can be explained by the following formulae.

First, signal propagation model $\mathbf{u}(= u_1, u_2, \ldots, u_{N+1})$ can be defined as Formula (1). In Formula (1), $\delta_i^{0/1}$ represents the difference of signal propagation delays for the ith selector unit when the challenge is zero or one.

$$\begin{cases} u_1 = \frac{\delta_1^0 - \delta_1^1}{2} \\ u_i = \frac{\delta_{i-1}^0 - \delta_{i-1}^1 + \delta_i^0 - \delta_i^1}{2} & (i = 2, \ldots, N) \\ u_{N+1} = \frac{\delta_N^0 + \delta_N^1}{2} \end{cases} \tag{1}$$

Next, challenge model $\mathbf{v}\ (= v_1, v_2, \ldots, v_{N+1})$ can be defined as Formula (2). In this formula, $\mathbf{C}\ (= C_1, C_2, \ldots, C_N)$ represents the challenge.

$$\begin{cases} v_l(\mathbf{C}) = \prod_{i=l}^{N} (1 - 2C_i) & (l = 1, \ldots, N) \\ v_{N+1}(\mathbf{C}) = 1 \end{cases} \tag{2}$$

Then, the difference Δ of signal propagation delays provided to the arbiter circuit can be represented by using the signal propagation model **u** and the challenge model **v**. This can be defined as Formula (3).

$$\Delta = \mathbf{u}^\mathrm{T}\mathbf{v} \tag{3}$$

Hence, the response r can be determined as Formula (4). In an actual machine learning attack, both challenge model **v** and response r are provided to machine learning algorithm, and a learning model is generated.

$$r = \mathrm{sgn}(\Delta) \tag{4}$$

As mention above, since the arbiter PUF can be represented by the linear model, it is vulnerable against machine learning attacks. Therefore, the feed-forward arbiter PUF, the XOR arbiter PUF, and the lightweight PUF, which have the resistance against machine learning attacks, have been proposed as countermeasures [4, 12].

2.1.3 Feed-Forward Arbiter PUF [4]

The basic structure of the feed-forward arbiter PUF [4] is depicted in Fig. 2. The feed-forward arbiter PUF has the structure inserted feed-forward arbiters into the arbiter PUF. As shown in Fig. 2, the output of the feed-forward arbiter is used as the challenge bit. In this section, the feed-forward arbiter PUF, which has one feed-forward arbiter and 64 selector units, is explained as an example. First, in this feed-forward arbiter PUF, two signals from the 1st to the 56th selector unit are inputted to the feed-forward arbiter, and then an arrival order of signals is judged. Then, the output is provided to the 64th selector unit as the challenge bit. Thus, by inserting a non-linear structure (feed-forward arbiter) into the arbiter PUF, the feed-forward arbiter PUF can improve the resistance against machine learning attacks.

Fig. 2 Structure of feed-forward arbiter PUF

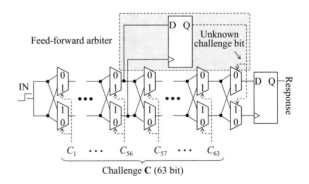

2.2 Related Works

In paper [13], the machine learning attack using side-channel information such as power consumption has been reported. This attack estimates the number of ones of the output for each arbiter PUF constituting the XOR arbiter PUF or lightweight PUF from a power consumption value. At this time, in the estimation of the number of ones for each arbiter PUF, the values such that all outputs are zeros or ones are used as 'good challenge' for the machine learning attack [13].

Also, a modeling attack using side-channel information for the feed-forward arbiter PUF has been reported [15]. This attack utilizes the difference of signal propagation delays with an unreliability response as side-channel information. The difference (side-channel information) is extracted by changing an environment setup such as temperature or supply voltage. Here, in modeling attacks, the learning model including unreliability responses causes decrease of the attack accuracy. By not using unreliability responses for the generation of the learning model, paper [15] improved the attack accuracy.

On the other hand, machine learning attacks using power consumption waveforms for the feed-forward arbiter PUF have not been studied.

3 Proposed Analysis Method

3.1 Outline of the Proposed Analysis Method

This study proposes a new power consumption aware machine learning attack for the feed-forward arbiter PUF. In the proposed analysis method, the hierarchical machine learning attack is performed. In this section, the proposed analysis method for the feed-forward arbiter PUF, which has 64 selector units and 1 feed-forward arbiter, is explained as an example. Figure 3 shows the concept of the proposed analysis method. The proposed analysis method consists of the first and the second machine learning attack (the first and the second stage). In the first stage, the machine learning attack for selector units from the 1st to the 56th is conducted, and then it predicts the output of the feed-forward arbiter. Next, in the second stage, the machine learning attack for all selector units is performed by utilizing the predicted output in the first stage. Finally, the response is predicted.

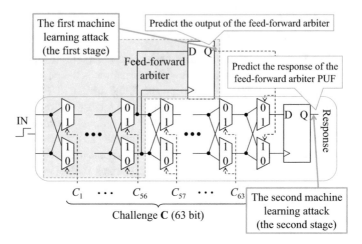

Fig. 3 Concept of the proposed analysis method

3.2 Learning Phase

The proposed analysis method consists of the learning phase to generate learning models and the analysis phase to predict the response. In the learning phase, learning models (learning model #1 and #2) are generated for the first and the second machine learning attack (the first and the second stage).

Figure 4 shows the generation method of the learning model #1. As shown in Fig. 4, the 56-bit challenge provided to 56 selector units (from the 1st to the 56th) and the output of the feed-forward arbiter are used for the generation of the learning model #1. Here, the output o of the feed-forward arbiter is an unknown value. Therefore, the proposed analysis method estimates the unknown value from the power consumption during the operation of the circuit. Specifically, DFF used as the arbiter circuit consumes large power due to the switching of data [13]. In other words, when the output o of D-FF is zero, the consumed power is low, and when the output o is one, the power is large. As a result, when the measured power consumption value is larger than a threshold value, the output o is estimated one, and when it is lower than the threshold value, the output o is estimated zero, as shown in Fig. 4 (i). Finally, the learning model #1 is generated.

In this machine learning, the 56-bit challenge $\mathbf{C}^{\#1}(= C_1, C_2, \ldots, C_{56})$ and the estimated value o are provided to the machine learning technique, as shown in Fig. 4 (ii). At this time, the challenge model $\mathbf{v}^{\#1}(= v_1^{\#1}, v_2^{\#1}, \ldots, v_{57}^{\#1})$ can be defined as Formula (5).

$$\begin{cases} v_l^{\#1}(\mathbf{C}^{\#1}) = \prod_{i=l}^{56}(1 - 2C_i) \quad (l = 1, \ldots, 56) \\ v_{57}^{\#1}(\mathbf{C}^{\#1}) = 1 \end{cases} \quad (5)$$

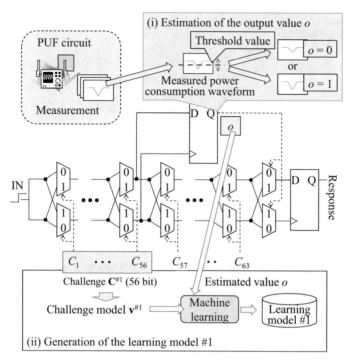

Fig. 4 Generation method of the learning model #1

Also, the signal propagation model $\mathbf{u}^{\#1}$ $\left(= u_1^{\#1},\ u_2^{\#1}, \ldots, u_{57}^{\#1} \right)$ can be defined as

$$
\begin{cases}
u_1^{\#1} = \frac{\delta_1^{\#1,0} - \delta_1^{\#1,1}}{2} \\
u_i^{\#1} = \frac{\delta_{i-1}^{\#1,0} - \delta_{i-1}^{\#1,1} + \delta_i^{\#1,0} - \delta_i^{\#1,1}}{2} \quad (i = 2, \ldots, 56), \\
u_{57}^{\#1} = \frac{\delta_{56}^{\#1,0} + \delta_{56}^{\#1,1}}{2}
\end{cases}
\tag{6}
$$

where $\delta_i^{\#1,\,0/1}$ represents the difference of signal propagation delays for the ith selector unit when the challenge is 0 or 1.

Hence, the estimated value o can be calculated by Formula (7).

$$
o = \text{sgn}\left(\mathbf{u}^{\#1\text{T}} \mathbf{v}^{\#1} \right)
\tag{7}
$$

Next, Fig. 5 shows the generation method of the learning model #2. As shown in Fig. 5, the 64-bit challenge $\mathbf{C}^{\#2}$ $(= C_1, C_2, \ldots, C_{63}, o)$ and the response r are used for the generation of the learning model #2. At this time, the value o, which is estimated in the generation of the learning model #1, is utilized as the 64th challenge

Fig. 5 Generation method
of the learning model #2

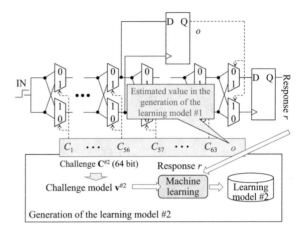

Generation of the learning model #2

bit (selection signal of the 64th selector unit). Consequently, the challenge model $\mathbf{v}^{\#2}\left(= v_1^{\#2}, v_2^{\#2}, \ldots, v_{65}^{\#2}\right)$ can be represented by Formula (8).

$$
\begin{cases}
v_l^{\#2}\left(\mathbf{C}^{\#2}\right) = \prod_{i=l}^{63} (1 - 2C_i) & (l = 1, \ldots, 63) \\
v_{64}^{\#2}\left(\mathbf{C}^{\#2}\right) = 1 - 2o \\
v_{65}^{\#2}\left(\mathbf{C}^{\#2}\right) = 1
\end{cases}
\tag{8}
$$

3.3 Analysis Phase

Figure 6 shows the outline of the analysis phase. The analysis phase performs the hierarchical machine learning attack for the feed-forward arbiter PUF, and then it predicts the response r' against the challenge C' for the test data.

First, to estimate the unknown output value o' of the feed-forward arbiter, the first machine learning attack is performed, as shown in Fig. 6 (i). This attack uses the generated learning model #1 in the learning phase (see Sect. 3.2) and the challenge model using the 56-bit challenge. Then, the output value o' is predicted.

Next, the second machine learning attack generates the challenge model using both predicted value o' and 63-bit challenge C'. Then, by using the challenge model and the generated learning model #2 (see Sect. 3.2), the second stage predicts the response r', as shown in Fig. 6 (ii).

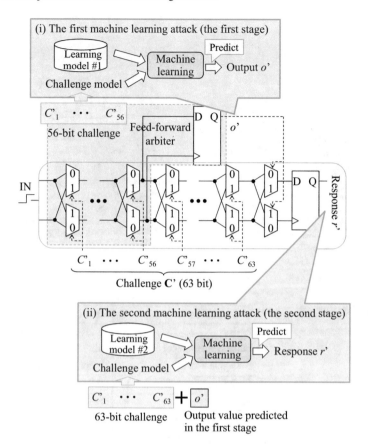

Fig. 6 Outline of the analysis phase

Finally, for the feed-forward arbiter PUF which has more than two feed-forward arbiters, in order to estimate outputs of arbiters easily from the measured power consumption, the challenge such that all outputs of all arbiters are zeros or ones is used. Also, the hierarchical machine learning attack for the PUF, which has k feed-forward arbiters, is performed $k + 1$ times to predict the response.

4 Experiments

4.1 Experimental Environment

Experiments used a side-channel attack standard evaluation board (SASEBO)-GII [16] and an oscilloscope. Figure 7 and Table 1 show the experimental environment.

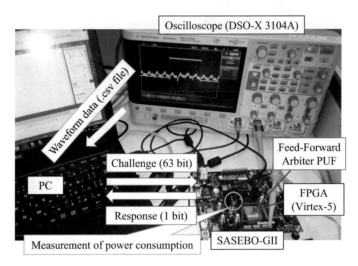

Fig. 7 Evaluation system

Table 1 Experimental condition

FPGA board	SASEBO-GII [16]
FPGA	Virtex-5 XC5VLX30
Implementation tool	Xilinx Design Suite 14.7
Floorplan	Xilinx PlanAhead v14.7
Oscilloscope	DSO-X 3104A
PUF	Feed-forward arbiter PUF
# of selector units	64
# of feed-forward arbiters	1
# of training data sets	1,000
# of test data sets	1,000
SVM	LibSVM
Kernel function	Linear kernel
SVM parameter	Default

The feed-forward arbiter PUF was designed by using the Verilog hardware description language (HDL), and it was implemented into a FPGA on SASEBO-GII. Here, the outline of the implemented feed-forward arbiter PUF is depicted in Fig. 8. The feed-forward arbiter PUF, which has 64 selector units and 1 feed-forward arbiter, was implemented, as shown in Fig. 8. Also, to measure the power consumption easily, additional toggle flip-flops (TFFs) were implemented to the feed-forward arbiter circuit.

Next, Fig. 9 shows the floorplan of the implemented feed-forward arbiter PUF. For the implementation, 64 selector units were placed from SLICE_X14Y76 to SLICE_X15Y13, as shown in Fig. 9. Then, the feed-forward arbiter was placed

Fig. 8 Outline of the implemented feed-forward arbiter PUF

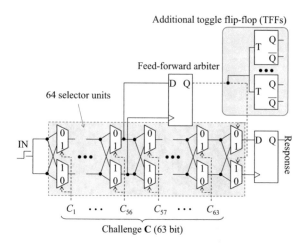

in SLICE_X16Y16. Also, for the measurement of the power consumption generated during the operation of the feed-forward arbiter PUF, the oscilloscope was used.

In the experiment, 1,000 random challenges were provided to the implemented feed-forward arbiter PUF, and then 1,000 responses and 1,000 power consumption waveforms were obtained as training data sets for the learning phase. Then, for the analysis phase, 1,000 random challenges were sent to the implemented feed-forward arbiter PUF, and 1,000 responses were received. Therefore, the experiment used 2,000 actual CRPs: 1,000 training data sets and 1,000 test data sets. Also, support vector machine (SVM) was used as a machine learning technique.

4.2 Experimental Results

Figure 10 shows the experimental result. In this figure, the vertical line shows the predicted rate, and the horizontal line shows the number of training data sets. As shown in Fig. 10, the predicted rate achieved above 90% with 1,000 training data sets. Therefore, the proposed analysis method is effective, and the feed-forward arbiter PUF is vulnerable against the proposed analysis method. Next, Fig. 11 shows the result of the first machine learning attack (the machine learning attack for the output of the feed-forward arbiter). As shown in Fig. 11, it could be confirmed that the predicted rate was above 95%.

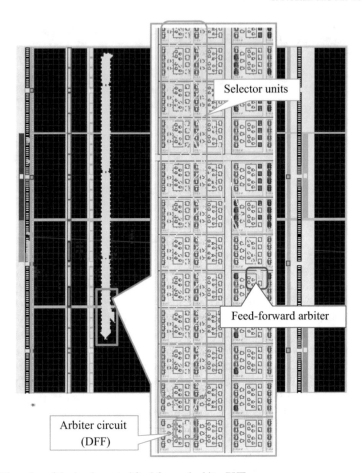

Fig. 9 Floorplan of the implemented feed-forward arbiter PUF

Fig. 10 Experimental result

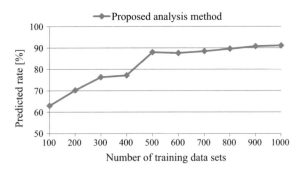

Fig. 11 Experimental result of the first machine learning attack

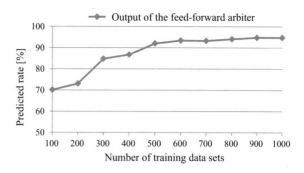

5   Conclusion

This study proposed a new power consumption aware machine learning attack for the feed-forward arbiter PUF. The proposed analysis method performs the hierarchical machine learning attack to predict the response. In the hierarchical machine learning attack, an intermediate output value of the feed-forward arbiter is estimated by the power consumption generated during the operation of the circuit. In the experiment using a FPGA board, the proposed analysis method could predict responses of above 90%. Hence, the validity of the proposed analysis method and the vulnerability of the feed-forward arbiter PUF were clarified.

Future works include the detailed evaluation of the feed-forward arbiter PUF which has more than two feed-forward arbiters and the countermeasure against the proposed analysis method.

Acknowledgements This paper is based on results obtained from a project commissioned by the New Energy and Industrial Technology Development Organization (NEDO).

References

1. ESIA: Over one million counterfeit semiconductors seized—ESIA supported customs operation with expertise. http://www.eusemiconductors.eu/images/static_website/newsroom/PR/E SIA_PR_JCO-Wafers_3Jul2017.pdf
2. Singh, S., Silakari, S.: An ensemble approach for cyber attack detection system: a generic framework. Int. J. Netw. Distrib. Comput. **2**(2), 78–90 (2014)
3. Yoshikawa, M., Sugioka, K., Nozaki, Y., Asahi, K.: Secure in-vehicle systems using authentication. Int. J. Netw. Distrib. Comput. **3**(3), 159–166 (2015)
4. Lee, J.-W., Lim, D., Gassend, B., Suh, G.E., Dijk, M.V., Debadas, S.: A technique to build a secret key in integrated circuits for identification and authentication applications. In: Proceedings of IEEE VLSI Circuits Symposium, pp. 176–179 (2004)
5. Suh, G.E., Devadas, S.: Physical unclonable functions for device authentication and secret key generation. In: Proceedings of 44th ACM/IEEE Design Automation Conference (DAC'07), pp. 9–14 (2007)
6. Shimizu, K., Suzuki, D.: Glitch PUF: extracting information from usually unwanted glitches. IEICE Trans. Fundamen. Electron. Commun. Comput. Sci. **E95-A**(1), 223–233 (2012)

7. Guajardo, J., Kumar, S.S., Schrijen, G.J., Tuyls, P.: FPGA intrinsic PUFs and their use for IP protection. In: Proceedings of 9th International Workshop on Cryptographic Hardware and Embedded Systems (CHES 2007), LNCS 4272, pp. 63–80. Springer (2007)

8. Hori, Y., Kang, H., Katashita, T., Satoh, A.: Pseudo-LFSR PUF: a compact, efficient and reliable physical unclonable function. In: Proceedings of IEEE 7th International Conference on ReConFigurable Computing and FPGAs (ReConFig'11), pp. 223–228 (2011)

9. Lim, D.: Extracting secret keys from integrated circuits. M.S. thesis, MIT (2004)

10. Rührmair, U., Sölter, J., Sehnke, F., Xu, X., Mahmoud, A., Stoyanova, V., Dror, G., Schmidhuber, J., Burleson, W., Devadas, S.: PUF modeling attacks on simulated and silicon data. IEEE Trans. Inf. Forensics Secur. **8**(11), 1876–1891 (2013)

11. Majzoobi, M., Koushanfar, F., Potkonjak, M.: Testing techniques for hardware security. In: Proceedings of IEEE International Test Conference (ITC 2008), pp. 1–10 (2008)

12. Majzoobi, M., Koushanfar, F., Potkonjak, M.: Lightweight secure PUFs. In: Proceedings of IEEE/ACM International Conference on Computer Aided Design (ICCAD), pp. 670–673 (2008)

13. Mahmoud, A., Rührmair, U., Majzoobi, M., Koushanfar, F.: Combined modeling and side channel attacks on strong PUFs. IACR Cryptology ePrint Archive: Report 2013/632 (2013)

14. Nozaki, Y., Yoshikawa, M.: EM based machine learning attack for XOR arbiter PUF. In: Proceedings of the 2017 Asia Conference on Machine Learning and Computing (ACMLC 2017), pp. 140–144 (2017)

15. Kumar, R., Burleson, W.: Side-channel assisted modeling attacks on feed-forward arbiter PUFs using silicon data. In: Proceedings of RFIDSec 2015, LNCS 9440, pp. 53–67. Springer (2015)

16. Research Center for Information Security: Evaluation Environment for Side-channel Attacks. National Institute of Advanced Industrial Science and Technology, http://www.risec.aist.go.jp/project/sasebo/

Adaptive Generative Initialization in Transfer Learning

Wenjun Bai, Changqin Quan and Zhi-Wei Luo

Abstract In spite of numerous researches on transfer learning, the consensus on the optimal method in transfer learning has not been reached. To render a unified theoretical understanding of transfer learning, we rephrase the crux of transfer learning as pursuing the optimal initialisation in facilitating the to-be-transferred task. Hence, to obtain an ideal initialisation, we propose a novel initialisation technique, i.e., adapted generative initialisation. Not limit to boost the task transfer, more importantly, the proposed initialisation can also bound the transfer benefits in defending the devastating negative transfer. At first stage in our proposed initialisation, the in-congruency between a task and its assigned learner (model) can be alleviated through feeding the knowledge of the target learner to train the source learner, whereas the later generative stage ensures the adapted initialisation can be properly produced to the target learner. The superiority of our proposed initialisation over conventional neural network based approaches was validated in our preliminary experiment on MNIST dataset.

Keywords Transfer learning · Negative transfer · Initialisation · Bayesian neural network

W. Bai (✉) · C. Quan · Z.-W. Luo
Department of Computational Science, Kobe University, 1-1, Rokkodai,
Nada, Kobe 657-8501, Japan
e-mail: bwj@cs11.cs.kobe-u.ac.jp

C. Quan
e-mail: quanchqin@gold.kobe-u.ac.jp

Z.-W. Luo
e-mail: luo@gold.kobe-u.ac.jp

© Springer Nature Switzerland AG 2019
R. Lee (ed.), *Computer and Information Science*, Studies in Computational
Intelligence 791, https://doi.org/10.1007/978-3-319-98693-7_5

1 Introduction

1.1 Transfer Learning as Initialisation

Recalling the over-arching goal of transfer learning is to facilitate a novel learning process via using obtained knowledge in previous solved tasks. However, despite of fruitful researches on transfer learning, the agreement on the optimal surrogates and manners of transfer learning has not been made. The mainstream approaches range from the shared training samples [1], the shared representations [2], to the reused model parameters [3]. To provide a common theoretical grounding in discussing transfer learning, we turn our direction to transfer learning in humans.

In contrast to the machines, since toddlers, humans are wired with the problem solving skill that is able to distill previous task knowledge. Irrespective of how novel the incoming tasks are, to human task solvers, the learning is barely initiated at scratch. Inspired from this, we speculate the crux of transfer learning is on finding optimal initialisations in solving the target task.

In a nutshell, all transfer learning can be perceived as the pursuing of optimal initialisations in training the target learner.[1] Initialised target learners are then fine-tuned—where there are enough training samples in the target task—to further improve the task performance. Different to previous attempts in finding the optimal initialisation for a single learner, the aim in transfer learning is to find the initialisations that is suitable for the target learner.

To design an initialisation technique that is tailored specific for multiple learners in transfer learning, we propose a novel initialisation technique: **adapted generative initialisation**. In this novel initialisation technique, two computational steps are sequentially executed, i.e., adaption and generative stages. The primary stage, i.e., adaptation, is responsible for incorporating the learner divergences between the source and target learner. The follow-up generative stage ensures the fed initialisation to be dimensionality compatible with the target learner.

1.2 Initialisation to Defend Negative Transfer

Besides the direct benefit in improving the (inductive) transfer learning, our approach can defend the attacks from a special but detrimental form of transfer learning: negative transfer. Negative transfer—a term borrowed from cognitive science—occurs in a situation where the prior learning of a source task interferes with the later learning of a target task. To formalise our discussion on negative transfer, considering two sequentially to-be-learned tasks with varying degrees of complexity; e.g., T1 (a complex task) and T2 (a simple task), we then assign two corresponding learners,

---

[1]A learner in this article represents any types of computational models, such as neural networks.

e.g., M1 (a cumbersome learner) and M2 (a compact learner) to learn the foregoing tasks. Dependent upon the congruence on task and learner complexity and the learning sequence, there are four distinctive transfer learning cases, i.e., T1D1 $\rightarrow$ T2M2; T1M2 $\rightarrow$ T2M1; T2M1 $\rightarrow$ T1M2; T2M2 $\rightarrow$ T1M1.[2] Under this finer specification of transfer learning, we speculate the real culprits of negative transfer are the incongruence on task and learner complexity, and the ordering of learning. Based on this speculation, the proposed adapted generative initialisation is designed to defend the attacks from the negative transfer as it attenuates the model divergences and alleviates the learning order effect during the transfer.

The remaining article is structured as following: After a brief review on related works, we focus on demonstrating our proposed adapted generative initialisation and its two inner stages, i.e., adaptation and generative stages in detail. We then extend our implementation from original dual learners to multiple learners (learners $\geq$ 3). To evaluate the proposed initialisation, we pit it against conventional approaches on MNIST dataset to see whether or not our proposed initialisation is superior in facilitating the 'positive' transfer learning and defending the negative transfer.

Our contributions in this work should be recognised as: (1) To authors' knowledge, it is first time the transfer learning is framed as the initialisation issue. (2) To obtain an optimal initialisation for inductive transfer learning, we propose a novel adapted generative initialisation. (3) Through a preliminary experiment, the proposed initialisation demonstrated the clear competitive edge over mainstream approaches in facilitating the 'positive' inductive transfer and circumventing the negative one.

2 Related Works

As a crucial branch in machine learning, numerous researches had been conducted in exploring the transfer learning. Among these attempts, several neural network based approaches needs to be credited here. In neural network based transfer learning, the extracted mid-level features are served as the knowledge to transfer. The production of these transferable features is accomplished by running forward propagation of a trained neural network [4]. To optimise the usage of extracted features in transfer learning, Hinton et al. [5] proposed a knowledge distillation technique to allow the compression of a cumbersome model to a compact one. The distilled knowledge, i.e., the cross entropy loss of a learned neural network, is augmented to fine-tune a to-be-learned neural network. However, both approaches are only applicable in 'normal' transfer learning but fail to sidestep a less studied but malign form of transfer learning: negative transfer. Previous attempts in defending the negative transfer emphasised

---

[2]$\rightarrow$ denotes the direction of knowledge transfer, e.g., T1D1 $\rightarrow$ T2D2 means the knowledge is extracted from a prior learning of a complex task through a cumbersome learner, then transferred to assist the learning of a simple task through a compact learner.

mostly on deriving the transfer bound via computing the task relatedness [6, 7], less attention was paid to the model-learner divergence in defending the negative transfer.

In speaking of the model initialisation, from early zero initialisation, random initialisation to late advanced glorot [8] and He initialisations [9], the breakthroughs in neural networks have always coupled with the development in more intricate initialisation methods. Both above-mentioned advanced initialisation techniques focus on finding the optimal layer wise number of neuron in each to-be-trained neural network. It has reached agreement that a proper initialisation is the key to smooth the model training and speed up the learning curriculum [10]. And, an optimal initialisation can also circumvent the issue of vanishing or exploding gradients [10] during the neural network training. However, these foregoing initialisations are only applicable in training the single learner, which is sub-optimal in dual or multiple learners. The initialisation for multiple learners, i.e., the situation we explore in this article, has rarely been studied.

3   Adapted Generative Initialisation

The proposed adapted generative initialisation is a Bayesian neural network(BNN) [11] based two-stage initialisation process, i.e., adaption and generative stage. As demonstrated in Fig. 1, the initial adaption stage (cf. ① in Fig. 1.) ensures a source BNN is initialised by the non-parametric bootstrapped [12] target BNN. In the sequential generative stage (cf. ② in Fig. 1.), it treats the variational inferred poste-

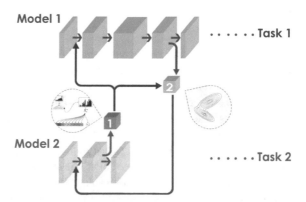

Fig. 1 Adapted Generative Initialisation. Figure 1 depicts the thumbnail of our proposed adapted generative initialisation. Two learners, i.e., model 1 and model 2, are assigned to learn two related tasks (inductive transfer learning). Two stages, i.e., adaption and generative stages, are plotted as red (cf. ①) and green (cf. ②) cubes. The computational processes within each stage are non-parametric bootstrapping and a Gaussian mixture model, respectively. Note here, the generative stage receives two input: one is from the previous adaptation stage, the other comes from the posterior parameters of model 1

rior parameters of a source BNN as an input to a generative model, e.g., a Gaussian mixture model (GMM). The employed GMM is then trained to generate the final transferrable weights in fine-tuning the target BNN. In practice, two GMMs are used separately in generating the transferrable weights for middle and classification layers of the target BNN.

3.1 Adaptation Stage

Unlike conventional initialisation techniques, which emphasised on the single learner, the adaptation stage in our initialisation technique is able to incorporate the divergence among learners. To achieve this goal, it demands to the knowledge in target learner to be distilled prior to the learning of the source task. However, the hurdle we need to jump off is to how to distill such knowledge before seeing any training samples for the target learner. The technique we resort on is the non-parametric bootstrapping [13]. This non-parametric bootstrapping method allows us to approximate the posterior parameters of the target learner. To formalise our discussion, considering to-be-approximate parameters are marked as $\theta_{bbt} = \{\theta_{bbt_1}, \ldots, \theta_{bbt_n}\}$, the non-parametric bootstrapping follows (1):

$$M \sim Geometric(p)$$

$$\pi \sim Dirichlet(0_1, \ldots, 0_M)$$

$$\mu_j \sim Uniform(-\infty, \infty) \tag{1}$$

$$k_j \sim Categorical(\pi)$$

$$\theta_{bbt} \sim \delta(\mu_{k_j})$$

The foregoing bootstrapping process can be described as: a to-be-approximated parameter (θ_{bbt}) comes from a specific value (μ_j) with a probability (π_j), but what these probabilities ($\pi_1, \ldots, \pi_M$) are due to the Dirichlet prior. The only part that remains unknown is how many values (M) is generated from. This is governed by the Geometric distribution where p can be seen as the probability that the current number of values (M) is the maximum number needed.

The intricacy of this adaption stage is not on this non-parametric bootstrapping process, but on the dual roles that these approximated parameters (θ_{bbt}) play. Their first role is to reserve it as one source of the input in feeding to later generative stage. The second role is to act as initialisation in training the source learner. Given enough training samples to train the source learner, i.e., D_s, and the foregoing posterior parameters as the initial value of the parameters, i.e., $\theta_{init_s} \leftarrow \theta_{bbt}$, relying on the variational inference, we are opted to minimise the description length, i.e., $\mathscr{F}$, to arrive our desired parameter: $\theta_s | D_s$, following:

$$\mathscr{F} = (L^N(\theta_s), D_s))|_{\theta \sim q(s)} + D_{KL}(q(\theta_s)||p(\theta_s)) \tag{2}$$

The source learner (network) loss can be written as the negative log probability of the data given the parameters,

$$L^N(\theta_s, D_s) = -\ln P(D_s|\theta_s) \tag{3}$$

and to resort on the mean field variational family in defining $q(\theta_s)$ [14], $q(\theta_s)$ can be rewritten as,

$$q(\theta_s) = \sum_{j=1}^{m} q(j) \tag{4}$$

Interestingly, the computation in this adaption stage, i.e., treating one's posterior as other's initialisation, coincides with the conventional parameter-in-transfer technique [3, 15] in transfer learning.

3.2 Generative Stage

In order to optimise the usage of previous obtained posterior parameters of the source learner, i.e., $\theta_s|D_s$, we then rely on a generative model to generate the needed initialisations to feed to the target learner. The chosen generative model is a Gaussian mixture model (GMM) [16]. To try out different generative model will be a interesting future research avenue. This generative model receives two input: one comes from the bootstrapped posterior parameters of the target learner, i.e., θ_{bbt}, the other comes from the derived parameter posterior of the source learner, i.e., $\theta_s|D_s$. Setting $K = 2$, we then map these inputs to two independent Gaussians. This degrades the problem of producing the optimal initialisation to GMM optimisation. These parameters can be defined as $\lambda = \{\pi_i, \mu_i, \Sigma_i\}$, where $i = (1, \ldots, M)$, here, μ_i, Σ_i, π_i means mean vectors, covariance matrices and the mixture weights from two component densities. Given the input, $X = (\theta_{bbt}, \theta_s|D_s)$, and the GMM model, i.e., $p(x) = \sum_{i=1}^{i} \pi_i N(x|\mu_i, \Sigma_i)$, we then resort on the expectation-maximization (EM) algorithm [17] to optimise these parameters, e.g., $\bar{\pi}_i, \bar{\mu}_i, \bar{\Sigma}_i$. The optimised parameters ensure the generation of layer-wise initialisation for the target learner (θ_t) via $\theta_t \sim N(\bar{\pi}_i, \bar{\mu}_i, \bar{\Sigma}_i)$.

To render two stages together, the overall pseudocode of our proposed adapted generative initialisation is presented below as Algorithm 1.

Algorithm 1 Adapated Generative Initialisation

INPUT: D_s: the training dataset in the source task, M: the number of bootstrapping points
OUTPUT: θ_t: the transferred parameters for the target learner

1: **procedure** ADAPTED GENERATIVE INIT(D_s; M)
2:    **function** *generative stage*(D_s)
3:      **for** $d_s \in D_s$ **do**
4:        **Initialise:** θ_{init_s} : initialise the parameters in the source learner
5:        **function** *adapation stage*(M)
6:          **for** $m \in M$ **do**
7:            θ_{bb_t}: non-parametric bootstrapped on the target model, following the Eq. (1)
8:            $\theta_{init_s} \leftarrow \theta_{bb_t}$
9:          **end for**
10:          **return** θ_{init_s}; θ_{bb_t}
11:        **end function**
12:        $\theta_s|D_s \leftarrow (\theta_s|\theta_{init_s}, d_s)$: inferred parameters in the source model based on training samples and bootstrapped parameters from the target model, following the Eqs. (2) to (4)
13:      **end for**
14:      **return** $\theta_s|D_s$; θ_{bbt}
15:      **function** GMM(X)
16:        $X \leftarrow concatenate(\theta_{bb_t}; \theta_s|D_s)$
17:        **for** $x \in X$ **do**
18:          $\theta_t \leftarrow GMM(x)$: generate the transferrable parameters in a trained GMM
19:        **end for**
20:        **return** θ_t
21:      **end function**
22:      **return** θ_t
23:    **end function**
24: **end procedure**

3.3 Multiple Learners

The proposed initialisation technique can easily extend to the case of multiple learners, i.e., where multiple target and source learner(s) are assigned to achieve the transfer learning. Considering a case of multiple learners, where the first two tasks are served as source tasks, leaving the final task as the target task, there are two variations in applying our proposed initialisation technique here. The sketches on these two variations are plotted below in Figs. 2 and 3.

From Figs. 2 and 3, it is clear that two variations in implementing our proposed initialisation differ in the learning order of the two source tasks, i.e., concurrent or sequentially learned. Irrespective of the sequential order in learning, the implemented adapted generative initialisation can always guarantee the produced initialisations to contain the knowledge from all learners.

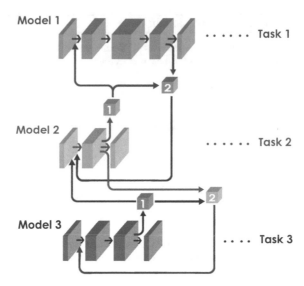

Fig. 2 Variation 1 of multiple learners. In this variation, we assume three learners are trained sequentially, i.e., model 1 → model 2 → model 3. The training of model 3 will not initiate until the finishing of the model 2. The implementation of adapted generative initialisation in transfer from model 2 → model 3 is exactly the same as model 1 → model 2, which is described in Fig. 1

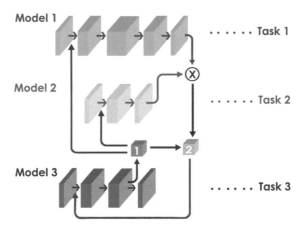

Fig. 3 Variation 2 of multiple learners. In this variation, we assume the two source tasks, e.g., task 1 and task 2 are learned concurrently. Hence, the bootstrapped parameters, which are yielded from the adaption stage (cf. ①) will be fed to both model 1 and model 2 (two brown upward pointing arrows). The trained features from model 1 and model 2 are then bilinear pooled (cf. ⓧ) [18] to serve as the second input to the generative stage

4 Experiment on MNIST

To validate our proposed initialisation, i.e., adapted generative initialisation, we implemented it on the MNIST dataset. To demonstrate the harming effect of negative transfer and the futile defence of conventional approaches, we artificially constructed two tasks with their varying degrees of difficulties, i.e., $T1 \& T2$.

4.1 Dataset and Model

The original MNIST dataset is composed of 60,000 single label annotated images of handwritten digits. To tailor this dataset for our experimental purpose, i.e., constructing a source and target task, respectively, we unevenly split the dataset in making one multi-class and binary classification task. In doing so, we intentionally create two tasks with varying levels of difficulties.

To demonstrate the effect of negative transfer, we also assign two learners with differentiated complexities, i.e $M1 \& M2$, to learn foregoing tasks. This consists a Latin square like 4 transfer learning situations (cf. Table 2). The model-wise configurations are delineated in Table 1. We pit our approach(initialisation) and two mainstream neural network based transfer learning techniques, e.g., mid-level feature representation and knowledge distillation, against their corresponded non-transfer benchmarks. We operationally define the 'positive' transfer as the elevation in testing accuracy in comparison to the non-transfer benchmark. Besides the 'positive' transfer effect, to observe whether the transfer technique successfully alleviate the attacks from negative transfer is our second objective. The occurrence of negative transfer is then defined as attenuated transfer performance, i.e., the lowered testing accuracy rate in comparison to the non-transfer benchmarks.

4.2 Results

From the results that are presented in Table 2 and Fig. 4, it is clear that the conventional approaches, e.g., mid-level feature extraction and knowledge distillation, failed to guarantee the success of transfer in diversified transfer circumstances. When the

Table 1 Model configurations

Model index	Model structure	# of parameters
M1	Input $\rightarrow$ Dense(100) $\rightarrow$ Dropout(0.3) $\rightarrow$ Dense(50) $\rightarrow$ Dense(0.3) $\rightarrow$ Dense(25) $\rightarrow$ Softmax	85,033
M2	Dense(20) $\rightarrow$ Softmax	1,570

$\rightarrow$ presents the flow of the data, and in-bracket number indexes the number of neurons used in each layer. Note here, the Bayesian neural networks that are used in the adapted generative initialisation follows the same layer-wise configuration with additional spheric Gaussian priors for the parameters

Table 2 Comparison of different inductive transfer approaches on MNIST

Transfer from	T1M1	T2M1	T1M2	T2M2
Transfer to	T2M2	T1M2	T2M1	T1M1
Transfer means	Testing accuracy (%)			
Mid-level feature extraction	94.00 ± 0.56	68.15 ± 1.10	95.76 ± 0.14	33.60 ± 3.31
Knowledge distillation	**98.25 ± 0.09**	**95.13 ± 0.07**	98.19 ± 0.01	94.97 ± 0.23
Without transfer NN	97.03 ± 0.02	95.06 ± 0.01	98.68 ± 0.02	97.65 ± 0.03
Adapted generative initialisation	**98.48 ± 0.08**	**94.70 ± 0.20**	**97.52 ± 0.09**	**93.03 ± 0.11**
Without transfer Bayesian NN	97.68 ± 0.02	94.69 ± 0.01	97.32 ± 0.01	88.97 ± 0.02

In this experiment, T_1: a multiclass classification task (MNIST 0-7); T_2: a binary classification task (MNIST 8-9); M_1: a relative deep neural network (five hidden layers); M_2: a relative shallow neural network (one hidden layer). We ran each technique 10 times on above-mentioned four cases. The negative transfer is defined as the attenuated performance on the transferred target task compares to the benchmark, i.e., the learning without transfer. As running of our proposed adapted initialisation relies on Bayesian neural networks, it demands different benchmarks other than previous non-transfer neural network ones

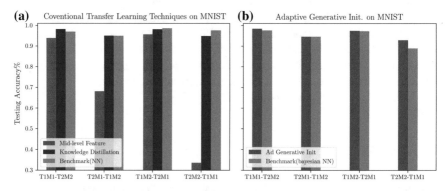

Fig. 4 Empirical Results on MNIST. In **a**, we pit two conventional approaches, i.e., mid-level feature extraction [4] and knowledge distillation [5] against the non-transfer neural networks, whereas in **b**, we also compare our approach with the non-transfer Bayesian neural network

divergence between tasks and learners is small with descended task difficulties, i.e., from T1M1 → T2M2 in our experiment, conventional approaches are able to bound the transfer effect; whereas in other situations (except knowledge distillation in T2M1 → T1M2), both approaches are weak in facilitating the 'positive' transfer and further defending the negative transfer. On the contrary, the robustness of our initialisation technique is observed as its ensured 'positive' transfer effects in all four training situations.

5 Conclusions

To deepen our understanding of transfer learning, in this pioneer research, we initiate a novel view that the goal of transfer learning can be framed as the pursuit of an optimal initialisation. Following this view, we propose a new initialisation technique, i.e., adapted generative initialisation, to incorporate the discrepancy between target and source learner(s) in producing the optimal initialisation for the target learner. Moreover, rather than facilitating the mere 'positive' transfer, our approach excels in sidestepping the negative transfer, in which the previous learning interferes with current learning process.

Through an empirical experiment on MNIST, in which we pit our approach against several mainstream neural network based transfer techniques, our technique proved its success in bounding the 'positive' transfer and defending the negative transfer. The observed superiority of our initialisation technique unequivocally provides a glimpse of future in better initialised transfer learning.

Acknowledgements This study is partially supported by the Okawa Foundation for Information and Telecommunications, and National Natural Science Foundation of China under Grant No. 61472117.

References

1. Jiang, J., Zhai, C.: Instance weighting for domain adaptation in NLP. In: Proceedings of the 45th Annual Meeting of the Association of Computational Linguistics, pp. 264–271 (2007)
2. Argyriou, A., Evgeniou, T., Pontil, M.: Multi-task feature learning. In: Advances in Neural Information Processing Systems, pp. 41–48 (2007)
3. Evgeniou, T., Pontil, M.: Regularized multi-task learning. In: Proceedings of the Tenth ACM SIGKDD International Conference on Knowledge Discovery and Data Mining, pp. 109–117. ACM (2004)
4. Oquab, M., Bottou, L., Laptev, I., Sivic, J.: Learning and transferring mid-level image representations using convolutional neural networks. In: 2014 IEEE Conference on Computer Vision and Pattern Recognition (CVPR), pp. 1717–1724. IEEE (2014)
5. Hinton, G., Vinyals, O., Dean, J.: Distilling the Knowledge in a Neural Network. http://arxiv.org/abs/1503.02531 (2015)
6. Lee, G., Yang, E., Hwang, S.J.: Asymmetric multi-task learning based on task relatedness and loss. In: Proceedings of the 33rd International Conference on International Conference on Machine Learning, ICML'16. JMLR.org, vol. 48, pp. 230–238. http://dl.acm.org/citation.cfm?id=3045390.3045416 (2016)
7. Mahmud, M.M., Ray, S.: Transfer learning using Kolmogorov complexity: basic theory and empirical evaluations. In: Advances in Neural Information Processing Systems, pp. 985–992 (2008)
8. Glorot, X., Bengio, Y.: Understanding the difficulty of training deep feedforward neural networks. In: Proceedings of the Thirteenth International Conference on Artificial Intelligence and Statistics, pp. 249–256 (2010)
9. He, K., Zhang, X., Ren, S., Sun, J.: Delving deep into rectifiers: surpassing human-level performance on ImageNet classification. In: Proceedings of the IEEE International Conference on Computer Vision, pp. 1026–1034 (2015)

10. Sutskever, I., Martens, J., Dahl, G., Hinton, G.: On the importance of initialization and momentum in deep learning. In: International Conference on Machine Learning, pp. 1139–1147 (2013)
11. Neal, R.M.: Bayesian Learning for Neural Networks, vol. 118. Springer Science & Business Media (2012)
12. Rubin, D.B.: The Bayesian Bootstrap, vol. 9, no. 1, pp. 130–134. https://projecteuclid.org/euclid.aos/1176345338 (1981)
13. Higgins, J.J.: Introduction to Modern Nonparametric Statistics (2003)
14. Blei, D.M., Kucukelbir, A., McAuliffe, J.D.: Variational inference: a review for statisticians. J. Am. Stat. Assoc. **112**(518), 859–877 (2017)
15. Bonilla, E.V., Chai, K.M., Williams, C.: Multi-task Gaussian process prediction. In: Advances in Neural Information Processing Systems, pp. 153–160 (2008)
16. Jacobs, R.A., Jordan, M.I., Nowlan, S.J., Hinton, G.E.: Adaptive mixtures of local experts. Neural Comput. **3**(1), 79–87 (1991)
17. Jordan, M.I., Jacobs, R.A.: Hierarchical mixtures of experts and the EM algorithm. Neural Comput. **6**(2), 181–214 (1994)
18. Lin, T.-Y., RoyChowdhury, A., Maji, S.: Bilinear CNN models for fine-grained visual recognition. In: Proceedings of the IEEE International Conference on Computer Vision, pp. 1449–1457 (2015)

A Multicriteria Group Decision Making Approach for Evaluating Renewable Power Generation Sources

Santoso Wibowo and Srimannarayana Grandhi

Abstract This paper formulates the renewable power generation sources' performance evaluation problem as a multicriteria group decision making problem, and presents a new multicriteria group decision making approach for effectively evaluating the performance of renewable power generation sources. The subjectiveness and imprecision of the decision making process is adequately handled by using intuitionistic fuzzy numbers. A multicriteria group decision making approach based on the TOPSIS approach and the degree of similarities are introduced for obtaining the relative degree of closeness value of each alternative on which the final decision can be made. An example is presented for demonstrating the applicability of the approach for dealing with renewable power generation sources' performance evaluation problem.

Keywords Decision makers · Renewable power generation sources
Performance evaluation · Multicriteria · Subjective assessments

1 Introduction

The continuous economic growth and rapid industrialization worldwide has led to a drastic increase in global energy consumption [1]. In addition to that, there are concerns about the environmental impact associated with the increase in high energy consumption [2].

As a result of this, government agencies around the world are focusing their attention on the use of renewable energy as an alternative energy source. Renewable energy is produced using natural resources that are constantly replaced and they never run out such as sunlight, wind and biomass [3]. Renewable energy sources are

S. Wibowo (✉) · S. Grandhi
School of Engineering & Technology, CQUniversity, Melbourne, Australia
e-mail: s.wibowo1@cqu.edu.au

S. Grandhi
e-mail: s.grandhi@cqu.edu.au

© Springer Nature Switzerland AG 2019
R. Lee (ed.), *Computer and Information Science*, Studies in Computational
Intelligence 791, https://doi.org/10.1007/978-3-319-98693-7_6

commonly accepted due to their abilities to reduce pollution and maintain environmental balance. In specific, these renewable energy sources provide a wide range of benefits including (a) the reduction of greenhouse gas emissions, (b) the improvement of energy self-sufficiency, (c) the creation of employment opportunities, and (d) the development of local economies [1, 4]. Due to the abilities of renewable energy sources to provide such benefits, the European Commission intends to increase the use of renewable energy sources by 27% which will help them achieve a 40% cut in greenhouse gas emissions by the year 2030 [5]. Thus, in order for government agencies to meet their expectations to be economically viable, environmentally friendly, and socially responsible, a structured approach is necessary for evaluating the performance of the most suitable renewable energy source for development and implementation.

Evaluating the performance of renewable power generation sources is challenging due to the availability of numerous alternatives, the presence of multiple decision makers, and the presence of subjectiveness and imprecision of the decision making process. As a result, a comprehensive evaluation of the renewable power generation source alternatives' overall performance is necessary.

Several research has been carried out in evaluating the performance of renewable power generation sources [6–8]. Ahmed and Tahar [6] present the analytical hierarchy process (AHP) approach for ranking renewable power generation sources in Malaysia. Based on this approach, multiple evaluation criteria are simultaneously considered. To reduce the cognitive burden on the decision maker, pairwise comparison is used for assessing the performance of individual alternatives and the relative importance of the criteria. This approach, however, the pairwise comparison process becomes cumbersome when the number of alternatives and criteria increases.

Georgopoulou et al. [7] present the elimination and et choice translating reality (ELECTRE) approach for evaluating the performance of renewable power generation alternatives. The approach integrates multiple evaluation criteria and categorizes the alternatives from the best ones to the less important ones in relation to the multiple evaluation criteria used. This approach, however, may lead to ranking irregularities when the alternatives appear to be very close to each other.

Streimikiene et al. [8] present the application of the technique for order preference by similarity to ideal solution (TOPSIS) approach for ranking renewable power generation technologies. Linguistic terms are used to assess the weights of all evaluation criteria and the performance rating of each alternative with respect to each criterion, leading to the development of a weighted fuzzy decision matrix. A closeness coefficient is defined for determining the ranking order of all renewable power generation technologies. This approach, however, is incapable of dealing with multiple decision makers.

To deal with the shortcomings of these approaches, this paper formulates the renewable power generation sources' performance evaluation problem as a multicriteria group decision making problem and presents a new multicriteria group decision making approach for effectively evaluating the performance of renewable power generation sources. The subjectiveness and imprecision of the decision making process is adequately handled by using intuitionistic fuzzy numbers. A multicriteria group decision making approach based on the TOPSIS approach and the degree of similarities are introduced for obtaining the relative degree of closeness value of each alternative on which the final decision can be made. An example is presented for demonstrating the applicability of the approach for dealing with renewable power generation sources' performance evaluation problem.

2   Performance Evaluation of Renewable Power Generation Sources

The performance evaluation of the renewable power generation sources involves several steps. These steps include (a) evaluating the performance ratings of available renewable power generation sources with respect to each criterion, and the relative importance of the evaluation criteria by the decision makers, (b) determining the criteria weighting and performance rating of renewable power generation sources, (c) aggregating the fuzzy criteria weightings and performance ratings for producing a weighted fuzzy performance matrix, and (d) calculating the relative degree of closeness for each renewable power generation source across all criteria [9].

Much research has been conducted in determining the relevant factors for evaluating the performance of renewable power generation sources in different perspectives [6, 8, 10–19] For example, Ahmad and Tahar [6] point out that deployment time and efficiency of the technology, capital cost, operational life and impact on environment factors should be addressed in the performance evaluation of renewable power generation sources. Streimikiene et al. [8] state that electricity supply availability, cost of grid connection, emission reduction and impact on human health are important factors in evaluating renewable power generation sources for development. Moura and de Almeida [10] believe that the total cost and the contribution of the technology are relevant factors in evaluating renewable power generation sources. Diakoulaki and Karangelis [11] point out that technology feasibility, operating cost and impact on environment factors are relevant for evaluating the performance of renewable generation source. Chatzimouratidis and Pilavachi [12] state that the total cost, minimization of CO_2 emissions, minimization of fuel, and risk should be included in the performance evaluation process of renewable power generation sources. Cristobal and Ramon [13] state that technology maturity, efficiency of the technology, capital cost, land requirement, and social and political acceptance are important factors for evaluating the performance of renewable power generation sources. Antunes et al. [14] point out that the total cost, environmental impacts associated with the new power plants and environmental costs of electricity generation must be taken into account while evaluating the performance of renewable power generation sources. Amer and Daim [15] believe that R&D cost, operation and maintenance cost, efficient of the technology, land requirement, and employment creation need to be considered in the performance evaluation process of renewable power generation sources. Meanwhile, Wang et al. [16] suggest that safety of energy system and employment creation are

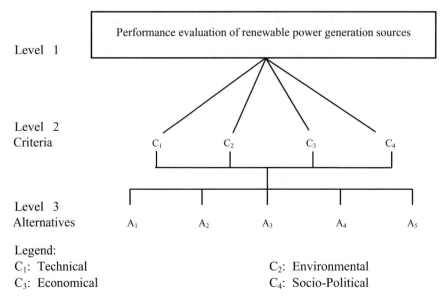

Legend:
C_1: Technical C_2: Environmental
C_3: Economical C_4: Socio-Political

$A_i (i = 1, 2, \ldots, n)$: Renewable power generation sources.

Fig. 1 The hierarchical structure of renewable power generation sources' performance evaluation problem

the relevant factors for assessing performance evaluation of renewable power generation sources. On top of that, Stein [17] believes that feasibility of the technology, compatibility with the national energy policy and impact on environment are crucial factors to be considered in the evaluation of renewable power generation sources. Brand et al. [18] claim that resource reserves, safety in covering peak demand, network stability, capital cost and disturbance of ecological balance are relevant factors for evaluating the performance of renewable power generation sources. At the same time, Pappas et al. [19] point out that emission reduction, need for waste disposal and, social and political acceptance should also be included in the performance evaluation process of renewable power generation sources. Troldborg et al. [20] believe that technology maturity, national economic development and impact on environment have to be taken into account while evaluating the performance of renewable power generation sources.

Based on the literature review presented above, four important criteria for evaluating the performance of renewable power generation sources are determined. The four criteria include (a) Technical (C_1), (b) Environmental (C_2), (c) Economical (C_3), and (d) Socio-Political (C_4). Figure 1 shows the hierarchical structure of renewable power generation sources performance evaluation problem.

Technical (C_1) criteria focus on the technical capabilities of the renewable power generation sources to meet or exceed the expectations of the customers. This criteria is assessed by technology maturity, efficiency, reliability, dependability, safety of the energy system, and network stability [6, 19].

Environmental (C_2) criteria reflect on the impact of renewable energy sources on environment. Al Garni et al. [21] highlight the undesirable effects of non-renewable energy sources and their role in pollution and global warming. In order to reduce carbon footprint, it is important to consider the environmental friendliness of renewable energy sources [7]. This is assessed by the impact on environment, need for waste disposal, emission reduction, and land requirement [11, 21].

Economical (C_3) criteria reflect on the economic effect in the performance evaluation of the renewable power generation source to the organization [8]. This criteria reflect on the ability to offer a cost-effective renewable energy alternative. The costs associated with the renewable energy sources include capital cost, cost of grid connection, and operation and maintenance costs [6, 14].

Socio-Political (C_4) criteria refer to both social and political factors. The issue of environmental protection and sustainability is influenced by both social attitudes and government policies. Hence, this criteria is assessed by both social and political acceptance of renewable energy alternative, its impact on nature and human health, employment opportunities it creates and its ability to meet government's energy policies [6, 18].

With the above identified criteria, each and every available renewable power generation source has to be evaluated for determining their overall performance across all the performance evaluation criteria so that the most appropriate renewable power generation source can be selected for development and implementation.

3 Multicriteria Group Decision Making Approach

Evaluating the performance of renewable power generation sources with respect to a set of evaluation criteria by multiple decision makers can be formulated as a multicriteria group decision making problem. This problem usually consists of a number of alternatives A (a_1, a_2, …, a_m) with respect to each criterion C (c_1, c_2, …, c_n) to be evaluated by multiple decision makers D (d_1, d_2, …, d_k).

Subjective assessments are often required for determining the performance of renewable power generation sources a_i with respect to each criterion and the relative importance of the criteria. This is due to (a) incomplete information, (b) ambiguous information, and (c) subjective information [22, 23]. To adequately model the subjectiveness and imprecision of the decision making process, intuitionistic fuzzy numbers [24, 25] are used for representing the subjective assessments of individual decision makers. This is due to the effectiveness in tackling the subjectiveness and imprecision, and the simplicity for the decision makers to assign their subjective assessments in the form of a membership degree and a non-membership degree [26].

The steps involved in the multicriteria group decision making approach are described below:

Step 1: Construct the hierarchical structure for the decision problem to be solved.

Step 2: Obtain the performance rating of alternatives with respect to all available criteria from the decision makers. As a result, the decision matrix for each decision maker can be obtained as

$$
y_{ij}^k =
\begin{bmatrix}
\mu_{11}^k, \ v_{11}^k & \mu_{12}^k, \ v_{12}^k & \cdots & \mu_{1m}^k, \ v_{1m}^k \\
\mu_{21}^k, \ v_{21}^k & \mu_{22}^k, \ v_{22}^k & \cdots & \mu_{2m}^k, \ v_{2m}^k \\
\cdots & \cdots & \cdots & \cdots \\
\mu_{n1}^k, \ v_{n1}^k & \mu_{n2}^k, \ v_{n2}^k & \cdots & \mu_{nm}^k, \ v_{nm}^k
\end{bmatrix}
\tag{1}
$$

Step 3: Obtain the collective intuitionistic fuzzy performance values from the decision makers by averaging the fuzzy assessments made by individual decision makers as in (2).

$$
X =
\begin{bmatrix}
x_{11} & x_{12} & \cdots & x_{1m} \\
x_{21} & x_{22} & \cdots & x_{2m} \\
\cdots & \cdots & \cdots & \cdots \\
x_{n1} & x_{n2} & \cdots & x_{nm}
\end{bmatrix}
\tag{2}
$$

Step 4: Obtain the relative positive ideal solution $\alpha^+ = (\alpha_1^+, \alpha_2^+, \ldots, \alpha_n^+)$ of the criteria. The relative positive ideal value of the criterion c_j can be calculated by

$$
\alpha_j^+ =
\begin{cases}
\left\langle \max_{1 \le i \le m}\{\mu_{ij}\}, \ \min_{1 \le i \le m}\{v_{ij}\} \right\rangle = \left\langle \mu_j^+, v_j^+ \right\rangle, \ if \ C_j \in F_1 \\
\left\langle \min_{1 \le i \le m}\{\mu_{ij}\}, \ \max_{1 \le i \le m}\{v_{ij}\} \right\rangle = \left\langle \mu_j^+, v_j^+ \right\rangle, \ if \ C_j \in F_2
\end{cases}
\tag{3}
$$

where F_1 denotes the set of benefit criteria, F_2 denotes the set of cost criteria, and $1 \le j \le n$.

Step 5: Obtain the relative negative ideal solution $\alpha^- = (\alpha_1^-, \alpha_2^-, \ldots, \alpha_n^-)$ of the criteria. The relative negative ideal value $\alpha^- = (\alpha_1^-, \alpha_2^-, \ldots, \alpha_n^-)$ of the criterion c_j can be calculated by

$$
\alpha_j^- =
\begin{cases}
\left\langle \min_{1 \le i \le m}\{\mu_{ij}\}, \ \max_{1 \le i \le m}\{v_{ij}\} \right\rangle = \left\langle \mu_j^-, v_j^- \right\rangle, \ if \ C_j \in F_1 \\
\left\langle \max_{1 \le i \le m}\{\mu_{ij}\}, \ \min_{1 \le i \le m}\{v_{ij}\} \right\rangle = \left\langle \mu_j^-, v_j^- \right\rangle, \ if \ C_j \in F_2
\end{cases}
\tag{4}
$$

where F_1 denotes the set of benefit criteria, F_2 denotes the set of cost criteria, and $1 \le j \le n$.

Step 6: Determine the degree of indeterminacy π_j^+ of the relative positive ideal solution $\alpha_j^+ = (\mu_j^+, v_j^+)$ for each criterion c_j by using (5).

$$
\pi_j^+ = 1 - \mu_j^+ - v_j^+
\tag{5}
$$

where $\pi_j^+ \in [1, 0]$ and $1 \leq j \leq n$.

Step 7: Determine the degree of indeterminacy π_j^- of the relative positive ideal solution $\alpha_j^- = (\mu_j^-, v_j^-)$ for each criterion c_j by using (6).

$$\pi_j^- = 1 - \mu_j^- - v_j^- \tag{6}$$

where $\pi_j^- \in [1, 0]$ and $1 \leq j \leq n$.

Step 8: Calculate the degree of similarity g_{ij}^+ between intuitionistic fuzzy numbers $d_{ij} = (\mu_{ij}, v_{ij})$ of alternative a_i with respect to criterion c_j and the obtained relative positive ideal value α_j^+ of criterion c_j.

$$g_{ij}^+ = s(d_{ij}, \alpha_j^+) = 1 - \frac{\left|2(\mu_j^+ - \mu_{ij}) - (v_j^+ - v_{ij})\right|}{3} \times (1 - \frac{\pi_j^+ + \pi_{ij}}{2})$$
$$- \frac{\left|2(v_j^+ - v_{ij}) - (\mu_j^+ - \mu_{ij})\right|}{3} \times (\frac{\pi_j^+ + \pi_{ij}}{2}) \tag{7}$$

where $g_{ij}^+ \in [0, 1]$, $1 \leq i \leq m$, m is the number of alternatives and n is the number of criteria.

Step 9: Calculate the degree of similarity g_{ij}^- between intuitionistic fuzzy numbers $d_{ij} = (\mu_{ij}, v_{ij})$ of alternative a_i with respect to criterion c_j and the obtained relative positive ideal value α_j^- of criterion c_j.

$$g_{ij}^- = s(d_{ij}, \alpha_j^-) = 1 - \frac{\left|2(\mu_j^- - \mu_{ij}) - (v_j^- - v_{ij})\right|}{3} \times (1 - \frac{\pi_j^- + \pi_{ij}}{2})$$
$$- \frac{\left|2(v_j^- - v_{ij}) - (\mu_j^- - \mu_{ij})\right|}{3} \times (\frac{\pi_j^+ + \pi_{ij}}{2}) \tag{8}$$

where $g_{ij}^- \in [0, 1]$, $1 \leq i \leq m$, m is the number of alternatives and n is the number of criteria.

Step 10: Compute the weighted positive score S_i^+ and the weighted negative score S_i^- for each alternative a_i respectively.

$$S_i^+ = \sum_{j=1}^n w_j \times g_{ij}^+, \quad S_i^- = \sum_{j=1}^n w_j \times g_{ij}^- \tag{9}$$

where w_j is the weight of the criterion.

Step 11: Calculate the relative degree of closeness $T(a_i)$ of each alternative a_i by (10). Based on the relative degree of closeness $T(a_i)$, the ranking order of each alternative a_i can be determined. The alternative a_i with the largest $T(a_i)$ is the most preferred solution.

$$T(a_i) = \frac{S_i^+}{S_i^+ + S_i^-} \tag{10}$$

4 An Example

In this section, we apply the multicriteria group decision making approach presented above to solve the performance evaluation of renewable power generation sources problem.

The energy consumption in India has grown rapidly in the last decade due to liberal policies on foreign investments, making it the world's fourth-largest energy consumer [3]. The energy consumption in India is mostly dominated by non-renewable energy sources such as coal. However, there is huge potential for meeting energy demand through renewable energy sources such as hydro, wind, solar and biomas. The adoption of renewable energy is seen as a positive alternative to fulfil a number of national goals and at the same time, a timely effort for the central government in India to provide more affordable and reliable renewable energy.

In order to meet its energy needs through renewable sources, it is therefore important to select the most suitable renewable power generation source. The renewable power generation source performance evaluation process starts with the formation of a committee consisting of three decision makers. Four important criteria that are relevant for evaluating the performance of renewable power generation sources are used which include Technical (C_1), (b) Environmental (C_2), (c) Economical (C_3), and (d) Socio-Political (C_4). Five alternatives are presented which include solar PV (a_1), solar thermal (a_2), geothermal (a_3), wind (a_4), and biomass (a_5).

The steps involved in the performance evaluation process are described below.

Step 1: The hierarchical structure for the decision problem to be solved is constructed as in Fig. 1.

Step 2: The performance rating of alternatives with respect to all available criteria are obtained from the decision makers as shown in Tables 1, 2 and 3 respectively. Assume that the weights of the criteria for $C_1 = 0.25$, $C_2 = 0.3$, $C_3 = 0.15$, and $C_4 = 0.3$.

Step 3: The collective intuitionistic fuzzy decision matrix is obtained as in Table 4.

Table 1 Intuitionistic fuzzy decision matrix of decision maker D_1

	Criteria			
	C_1	C_2	C_3	C_4
a_1	(0.4, 0.6)	(0.5, 0.9)	(0.5, 0.7)	(0.4, 0.6)
a_2	(0.6, 0.7)	(0.5, 0.7)	(0.4, 0.6)	(0.1, 0.3)
a_3	(0.1, 0.3)	(0.1, 0.4)	(0.5, 0.8)	(0.3, 0.7)
a_4	(0.3, 0.7)	(0.5, 0.8)	(0.6, 0.8)	(0.5, 0.8)
a_5	(0.4, 0.8)	(0.3, 0.6)	(0.3, 0.5)	(0.3, 0.4)

Table 2 Intuitionistic fuzzy decision matrix of decision maker D_2

	Criteria			
	C_1	C_2	C_3	C_4
a_1	(0.2, 0.7)	(0.4, 0.7)	(0.3, 0.4)	(0.2, 0.5)
a_2	(0.3, 0.7)	(0.6, 0.9)	(0.5, 0.7)	(0.2, 0.6)
a_3	(0.1, 0.3)	(0.1, 0.5)	(0.5, 0.8)	(0.6, 0.7)
a_4	(0.4, 0.6)	(0.5, 0.8)	(0.1, 0.3)	(0.5, 0.7)
a_5	(0.2, 0.5)	(0.3, 0.6)	(0.4, 0.5)	(0.5, 0.9)

Table 3 Intuitionistic fuzzy decision matrix of decision maker D_3

	Criteria			
	C_1	C_2	C_3	C_4
a_1	(0.6, 0.8)	(0.3, 0.7)	(0.4, 0.7)	(0.3, 0.4)
a_2	(0.3, 0.4)	(0.4, 0.5)	(0.3, 0.8)	(0.6, 0.9)
a_3	(0.1, 0.3)	(0.1, 0.3)	(0.5, 0.8)	(0.3, 0.7)
a_4	(0.5, 0.9)	(0.2, 0.4)	(0.3, 0.7)	(0.7, 0.9)
a_5	(0.3, 0.5)	(0.3, 0.6)	(0.5, 0.8)	(0.4, 0.8)

Table 4 Collective intuitionistic fuzzy decision matrix

	Criteria			
	C_1	C_2	C_3	C_4
a_1	(0.4, 0.7)	(0.4, 0.8)	(0.4, 0.6)	(0.3, 0.5)
a_2	(0.4, 0.6)	(0.5, 0.7)	(0.4, 0.7)	(0.3, 0.6)
a_3	(0.1, 0.3)	(0.1, 0.4)	(0.5, 0.8)	(0.4, 0.7)
a_4	(0.4, 0.73)	(0.4, 0.67)	(0.33, 0.6)	(0.56, 0.8)
a_5	(0.3, 0.5)	(0.3, 0.6)	(0.4, 0.6)	(0.4, 0.7)

Step 4: The relative positive ideal solution of the criteria are obtained as

$$\alpha_1^+ = (0.6, 0.4), \alpha_2^+ = (0.5, 0.2), \alpha_3^+ = (0.4, 0.4), \alpha_4^+ = (0.5, 0.3).$$

Step 5: The relative negative ideal solution of the criteria are obtained as

$$\alpha_1^- = (0.3, 0.5), \alpha_2^- = (0.2, 0.4), \alpha_3^- = (0.1, 0.3), \alpha_4^- = (0.4, 0.7).$$

Step 6: The degree of indeterminacy π_j^+ of the relative positive ideal solution $\alpha_j^+ = (\mu_j^+, \nu_j^+)$ for each criterion c_j are determined as

$$\pi_1^+ = 0.2, \ \pi_2^+ = 0.1, \ \pi_3^+ = 0.3, \ \pi_4^+ = 0.1.$$

Step 7: The degree of indeterminacy π_j^- of the relative positive ideal solution $\alpha_j^- = (\mu_j^-, \nu_j^-)$ for each criterion c_j are determined as

Alternatives	Relative degree of closeness	Ranking
A_1	0.738	1
A_2	0.715	2
A_3	0.582	5
A_4	0.678	3
A_5	0.647	4

Table 5 The relative degree of closeness values and their rankings

$$\pi_1^- = 0.1, \ \pi_2^- = 0.2, \ \pi_3^- = 0.1, \ \pi_4^- = 0.2.$$

Step 8: The degree of similarity g_{ij}^+ between intuitionistic fuzzy numbers $d_{ij} = (\mu_{ij}, \nu_{ij})$ of alternative a_i with respect to criterion c_j and the obtained relative positive ideal value α_j^+ of criterion c_j are calculated as

$$g_{11}^+ = 0.874, \ g_{21}^+ = 0.795, \ g_{31}^+ = 0.829, \ g_{41}^+ = 0.853, \ g_{51}^+ = 0.913$$
$$g_{12}^+ = 0.728, \ g_{22}^+ = 0.903, \ g_{32}^+ = 0.794, \ g_{42}^+ = 0.873, \ g_{52}^+ = 0.785$$
$$g_{13}^+ = 0.676, \ g_{23}^+ = 0.638, \ g_{33}^+ = 0.527, \ g_{43}^+ = 0.621, \ g_{53}^+ = 0.538$$
$$g_{14}^+ = 0.644, \ g_{24}^+ = 0.727, \ g_{34}^+ = 0.806, \ g_{44}^+ = 0.624, \ g_{54}^+ = 0.713$$

Step 9: The degree of similarity g_{ij}^- between intuitionistic fuzzy numbers $d_{ij} = (\mu_{ij}, \nu_{ij})$ of alternative a_i with respect to criterion c_j and the obtained relative positive ideal value α_j^- of criterion c_j are calculated as

$$g_{11}^- = 0.711, \ g_{21}^- = 0.602, \ g_{31}^- = 0.662, \ g_{41}^- = 0.853, \ g_{51}^- = 0.913$$
$$g_{12}^- = 0.735, \ g_{22}^- = 0.914, \ g_{32}^- = 0.863, \ g_{42}^- = 0.784, \ g_{52}^- = 0.731$$
$$g_{13}^- = 0.862, \ g_{23}^- = 0.853, \ g_{33}^- = 0.849, \ g_{43}^- = 0.823, \ g_{53}^- = 0.6447$$
$$g_{14}^- = 0.755, \ g_{24}^- = 0.744, \ g_{34}^- = 0.761, \ g_{44}^- = 0.728, \ g_{54}^- = 0.729$$

Step 10: The weighted positive score S_i^+ and the weighted negative score S_i^- for each alternative a_i are calculated as

$$S_1^+ = 0.892, \ S_2^+ = 0.873, \ S_3^+ = 0.924, \ S_4^+ = 0.896, \ S_5^+ = 0.637.$$
$$S_1^- = 0.317, \ S_2^- = 0.348, \ S_3^- = 0.664, \ S_4^- = 0.424, \ S_5^- = 0.348.$$

Step 11: The relative degree of closeness $T(a_i)\kappa$ of each alternative a_1 can be calculated by (10). The alternatives are ranked in descending order. The alternative a_i with the largest relative degree of closeness $T(a_i)$ is the most preferred. Table 5 shows the calculated results. Based on the relative degree of closeness value obtained in Table 5, the ranking order of each alternative can be determined. Alternative a_1, solar PV, is the most suitable renewable power generation source for development and implementation as it has the highest relative degree of closeness value of 0.738.

5 Conclusion

Evaluating the performance of renewable power generation sources is challenging due to the availability of numerous alternatives, the presence of multiple decision makers, and the presence of subjectiveness and impreciseness of the decision making process. To effectively solve this problem, this paper has formulated the renewable power generation sources performance evaluation problem as a multicriteria group decision making problem and applied the multicriteria group decision making approach for evaluating the performance of renewable power generation sources. The approach is computationally simple, and its underlying concept is rational and comprehensible.

References

1. Luz, T., Moura, P., de Almeida, C.: Multi-objective power generation expansion planning with high penetration of renewables. Renew. Sustain. Energy Rev. **81**, 2637–2643 (2018)
2. Demirbas, A., Sahin-Demirbas, A., Hilal Demirbas, A.: Global energy sources, energy usage, and future developments. Energy Sources **26**, 191–204 (2004)
3. Balat, M., Ayar, G.: Biomass energy in the world, use of biomass and potential trends. Energy Sources **27**, 931–940 (2005)
4. Strantzali, E., Aravossis, K.: Decision making in renewable energy investments: a review". Renew. Sustain. Energy Rev. **55**, 885–898 (2016)
5. Blenkinsopp, T., Coles, S.R., Kirwan, K.: Renewable energy for rural communities in Maharashtra. India Energy Policy **60**, 192–199 (2013)
6. Ahmad, S., Tahar, R.M.: Selection of renewable energy sources for sustainable development of electricity generation system using analytic hierarchy process: a case of Malaysia. Renew. Energy **63**, 458–466 (2014)
7. Georgopoulou, E., Lalas, D., Papagiannakis, L.: A multicriteria decision aid approach for energy planning problems: the case of renewable energy option. Eur. J. Oper. Res. **103**, 38–54 (1997)
8. Streimikiene, D., Balezentis, T., Krisciukaitiene, I., Balezentis, A.: Prioritizing sustainable electricity production technologies: MCDM approach. Renew. Sustain. Energy Rev. **16**, 3302–3311 (2012)
9. Wibowo, S., Deng, H.: Consensus-based decision support for multicriteria group decision making. Comput. Ind. Eng. **66**, 625–633 (2013)
10. Moura, P.S., de Almeida, A.T.: Multi-objective optimization of a mixed renewable system with demand-side management. Renew. Sustain. Energy Rev. **14**, 1461–1468 (2010)
11. Diakoulaki, D., Karangelis, F.: Multi-criteria decision analysis and cost-benefit analysis of alternative scenarios for the power generation sector in Greece. Renew. Sustain. Energy Rev. **11**, 716–727 (2007)
12. Chatzimouratidis, A.I., Pilavachi, P.A.: Technological, economic and sustainability evaluation of power plants using the analytic hierarchy process. Energy Policy **37**, 778–787 (2009)
13. Cristobal, S., Ramon, J.: Multi-criteria Analysis in the Renewable Energy Industry. Springer, London (2012)
14. Antunes, C.H., Martins, A.G., Brito, I.S.: A multiple objective mixed integer linear programming model for power generation expansion planning. Energy **29**, 613–627 (2004)
15. Amer, M., Daim, T.U.: Selection of renewable energy technologies for a developing county: a case of Pakistan. Energy. Sustain. Dev. **15**, 420–435 (2011)

16. Wang, J.J., Jing, Y.Y., Zhang, C.F., Zhao, J.H.: Review on multi-criteria decision analysis aid in sustainable energy decision-making. Renew. Sustain. Energy Rev. **13**, 2263–2278 (2009)
17. Stein, E.W.: A comprehensive multi-criteria model to rank electric energy production technologies. Renew. Sustain. Energy Rev. **22**, 640–654 (2013)
18. Brand, B., Missaoui, R.: Multi-criteria analysis of electricity generation mix scenarios in Tunisia. Renew. Sustain. Energy Rev. **39**, 251–261 (2014)
19. Pappas, C., Karakosta, C., Marinakis, V., Psarras, J.: A comparison of electricity production technologies in terms of sustainable development. Energy Convers. Manag. **64**, 626–632 (2012)
20. Troldborg, M., Heslop, S., Hough, R.L.: Assessing the sustainability of renewable energy technologies using multi-criteria analysis: suitability of approach for national-scale assessments and associated uncertainties. Renew. Sustain. Energy Rev. **39**, 1173–1184 (2014)
21. Al Garni, H., Kassem, A., Awasthi, A., Komljenovic, D., Al-Haddad, K.: A multicriteria decision making approach for evaluating renewable power generation sources in Saudi Arabia. Sustain. Energy Technol. Assess. **16**, 137–150 (2016)
22. Wibowo, S., Deng, H.: Multi-criteria group decision making for evaluating the performance of e-waste recycling programs under uncertainty. Waste Manag. **40**, 127–135 (2015)
23. Wibowo, S., Grandhi, S.: Evaluating the performance of recoverable end-of-life products in the reverse supply chain. Int. J. Netw. Distrib. Comput. **5**, 71–79 (2017)
24. Atanassov, K.T.: Intuitionistic fuzzy sets. Fuzzy Sets Syst. **20**, 87–96 (1986)
25. Panwar, A.: Evaluation of kernel based Atanassov's intuitionistic fuzzy clustering for network forensics and intrusion detection. Int. J. Softw. Innov. **4**, 1–15 (2016)
26. Xu, Z.S., Yager, R.R.: Some geometric aggregation operators based on intuitionistic fuzzy sets. Int. J. Gen Syst. **35**, 417–433 (2006)

Incremental Singular Value Decomposition Using Extended Power Method

Sharad Gupta and Sudip Sanyal

Abstract In this paper we present a novel method to perform Incremental Singular Value Decomposition (ISVD) by using an adaptation of the power method for diagonalization of matrices. We find that the efficiency of the procedure depends on the initial values and present two alternative ways for initialization. Gram-Schmidt orthonormalization is employed to ensure the orthogonality of the generated singular vectors. The suggested procedure does not depend on any assumption regarding the nature of input matrix or the change (increment) in the input matrix. Moreover, the results do not deviate from exact values even after repeated increments to the original input matrix. In order to test the suggested technique we apply it to the task of Latent Semantic Indexing for query processing using the MEDLINE and TIME corpus. Seven hundred documents are used to build the initial matrix for MEDLINE and two hundred for TIME corpus. We then add one document at a time till the complete corpus is included. Our ISVD technique is applied each time a new document is added. The results obtained using both the methods for initialization are very similar to the exact results obtained using direct diagonalization routines. This implies that the proposed method remains stable even after hundreds of increments in the original matrix.

Keywords Gram-Schmidt orthonormalization · Incremental singular matrix decomposition · Latent semantic indexing · Power method

S. Gupta (✉)
Information Technology, Indian Institute of Information Technology, Allahabad, Allahabad, India
e-mail: sharad311285@gmail.com

S. Sanyal
Computer Science and Engineering, BML Munjal University, Gurgaon, India
e-mail: sudip.sanyal@bml.edu.in

© Springer Nature Switzerland AG 2019
R. Lee (ed.), *Computer and Information Science*, Studies in Computational Intelligence 791, https://doi.org/10.1007/978-3-319-98693-7_7

1 Introduction

Singular Value Decomposition (SVD) has been used for different tasks like low rank
approximation of matrix for rank reduction [4], face recognition using Principal
Component Analysis, image compression [6], Web searching for finding the correct
result for a given query using Latent Semantic Analysis [14], pseudo-inverse for the
non-square matrix [24] etc. SVD decomposes (usually a non-square) matrix into a
product of three matrices as given in (1) below.

SVD of a matrix $A_{m \times n}$ is given by the formula

$$A_{m \times n} = U_{m \times k} S_{k \times k} V_{k \times n}^T, \quad k = min(m, n) \tag{1}$$

In (1), U and V are orthogonal matrices.[1] The columns of U and V give a linear basis
for A's columns and rows respectively. V^T is the transpose of matrix V. Moreover,
S is a Diagonal matrix.

$$S = diag(s1, s2, \ldots, sk), \quad s1 > s2 > \cdots > sk \tag{2}$$

The classical methods for performing SVD is to assume that the matrix A is known
and execute standard algorithms as described by Golub and Van Loan [9] and Press
et al. [19]. In recent years the focus has shifted to incremental SVD where the matrix,
A, is not known a priori. These situations can arise when an initial matrix, A, is
available. However, with the arrival of new data the matrix, A, may change to a new
matrix, $\hat{A}$ such that the decomposition of $\hat{A}$ leads to $\hat{U}$, $\hat{S}$ and $\hat{V}$. Such situations may
arise, for example, when we perform Latent Semantic Analysis on some corpus and
the corpus changes with the arrival of new documents. This would lead to addition
of new rows (corresponding to new documents) and new columns (corresponding to
new terms) in A. The classical methods would attempt to perform the decomposition
using the new matrix, $\hat{A}$, starting from the initial steps. Incremental SVD, on the
other hand, employs the previously calculated values of the matrices U, S and V to
find the corresponding values for the new matrix, $\hat{A}$.

In this paper we present a fast incremental algorithm to perform singular value
decomposition of a matrix which is an adaptation of the power method [9, 23] for
diagonalization of matrices. The power method is a method that uses an iterative
improvement of the initial values. Thus, the efficiency of the method depends criti-
cally on the choice of the initial values. In this paper we present two possible methods
for initialization. In the first method (called method I) the initial values of $\hat{U}$ and $\hat{V}$
are simply equal to the U and V respectively with zero padding if required. However,
in the second method (called method II) we initialize $\hat{U}$ and $\hat{V}$ using the previously
calculated values of U and V and also employ the previous singular values S to
estimate the new singular values $\hat{S}$. Usually the power method gives only the first
eigenvector but it can be extended to calculate any number of eigenvectors.

---

[1]$UU^T = VV^T = I$, where I is an Identity Matrix.

The classical SVD is usually performed in $O(mn^2 + m^2n + n^3)$ time in the batch mode [4]. If we use Lanczos method then it takes $O(mnk^2)$ [4]. However, for dense low-rank matrices the time complexity reduces to $O(mnk)$ for incremental-SVD. We used power method for calculating the singular vectors which calculates the vector in an iterative manner. Each iteration has a time complexity of $O(mn)$. Thus, the overall time complexity depends on the number of iterations. Often this is considered to be a slow convergence method but if we approximate the starting assumption carefully then we may reduce the number of iterations as shown later in Sect. 5.

While the time complexity is an important issue in techniques for incremental learning, the accuracy of the results is equally important. Ideally, we would want the results to remain accurate even after a number of increments i.e. a number of changes in the initial matrix A. The existing approaches to incremental SVD suffer from a steady divergence of the output from the exact results as the input matrix undergoes repeated increments [20]. Thus, the existing approaches use one round of the classical batch mode SVD after a few increments of the matrix, A. The main advantage of our approach is not the reduction of time complexity but its sustainability for long term i.e. the singular values and corresponding singular vectors are quite accurate even after repeated increments and we need not run the batch mode SVD.

Some other decompositions have been used to calculate the rank of a matrix. These methods also offer easy updates as the input matrix changes. One of such decomposition is the URV-decomposition given by Stewart [26] which has been used for subspace tracking [27]. However, it can not provide Singular Values which are needed in techniques like Latent Semantic Indexing (LSI) [8, 11]. Even though SVD is a special case of URV decomposition, since the latter does not provide singular values so we can not use it for applications where these are required.

The proposed approach has been tested on the task of Latent Semantic Indexing (LSI) for measuring document similarity. Standard benchmark corpus has been used and the results obtained using our incremental approach has been compared with the results obtained using direct diagonalization methods. The precision and recall obtained by our approach is very similar to the one obtained using the direct approach.

We have discussed some previous work on incremental SVD in Sect. 2. Our approach along with its application on Latent Semantic indexing (LSI) has been discussed in Sects. 3 and 4. Results of our method are presented in Sect. 5 along with their comparison with previous method. We present our concluding remarks in Sect. 6.

2 Related Work

There have been several approaches to incremental SVD in the past based on different assumptions regarding the input matrix and the type of increment [4, 6, 12, 13, 16, 20, 25, 30]. Some previous approaches like Rehurek assume the input matrix to be a sparse matrix [21] while others assume it to be a dense matrix [3, 16, 30]. Some previous works assume that only new rows or columns will be added

[16, 21, 29] while others allow addition of both rows and columns [3, 30]. Distributed incremental SVD algorithm has also been proposed [21].

Zha and Simon present a method for updating SVD but it works efficiently only for input matrices that are dense [30]. Chandrasekaran et al. propose a method for performing SVD incrementally but it is limited to approximating a single vector at a time and because of that there is loss of orthogonality [6]. Similarly, the method given by Levy and Lindenbaum is also vulnerable to loss of orthogonality [16]. An updating SVD algorithm for repeated or "close" singular values is also presented by Davies and Smith [7].

There are some methods which work for a few increments in the original matrix [12]. However, each increment produces a small divergence from the exact value and after a few increments the difference between the exact result and the result produced by the incremental algorithm becomes quite large [20]. These algorithms work well for small changes in the matrix, A, but for large changes in the input matrix, we have to run the whole system as batch mode SVD. Some other incremental methods for computing the low-rank SVD of a matrix have been presented by earlier researchers [1, 4].

Low rank approximation method for SVD have also presented by various researchers [1, 4, 28]. These methods calculate a low rank SVD of the given new matrix instead of computing the SVD of the whole matrix. On the other hand, the method presented in the present work can perform both the operations i.e. we can calculate a low rank SVD of the updated matrix or an SVD of the full matrix as per the requirements of the application.

As stated earlier, the present work uses the Power Method for updating Singular Value Decomposition which calculates the Eigenvectors and Eigenvalues of a matrix. There are various approaches for updating Eigenvalues and Eigenvectors [2, 5, 10, 15, 17, 22]. Method given by Hall et al. introduces a "forgetting factor" to estimate deviations from exact values [10]. This method was subsequently employed by Oyama et al. and extended to give an incremental method for simple Principal Component Analysis (PCA) [17].

Balsubramani et al. presented a convergence based method which depends on two parameters: the first is the convergence rate and the second is a multiplier. The final convergence rate depends on these two parameters. They have also used Gram-Schmidt orthogonalization for ensuring the orthogonality of Eigenvectors. QR decomposition based method has been used by Kwok [15] for performing SVD incrementally to calculate the k-rank SVD of a matrix.

Incremental PCA algorithm for online learning algorithm has also been proposed by Ozawa et al. [18]. They only update the Eigenvalues and Eigenvectors whenever they find that the previous eigenspace can not accurately represent the new data. This decision is based on a threshold and it needs to be changed. An SVD based PCA update algorithm has also been proposed by Zhao et al. [31] which uses rotations to find the orthogonal vectors. The methods proposed in the present work do not suffer from these problems and do not make any assumptions about the nature of the input matrix or the type of increments that we may make for the input matrix.

3   Incremental Singular Value Decomposition

Let $A_{m \times n}$ be the input matrix whose SVD is given by Eq. (1). We can update it in the following ways:

- Increment in number of columns: Let $R_{m \times p}$ be the new columns added to the old matrix then the new matrix will be

$$\hat{A}_{m \times (n+p)} = \left(A_{m \times n} \;\; R_{m \times p} \right)$$

- Increment in number of rows: Let $T_{q \times n}$ be the new rows added to the matrix then the new matrix will be

$$\hat{A}_{(m+q) \times n} = \left(\begin{array}{c} A_{m \times n} \\ T_{q \times n} \end{array} \right)$$

- Deleting a column: It can be achieved by simply putting all elements of the column to zero with no change in the size of matrix

$$\hat{A}_{m \times n} = \left(X_{m \times t} \;\; 0_{m \times 1} \;\; Y_{m \times (n-(t+1))} \right)$$

- Deleting a row: It can be achieved by simply putting all elements of the row to zero with no change in the size of matrix

$$\hat{A}_{m \times n} = \left(\begin{array}{c} X_{t \times n} \\ 0_{1 \times n} \\ Y_{(m-(t+1)) \times n} \end{array} \right)$$

Let us consider an incremented matrix $\hat{A}_{(m+p) \times (n+q)}$ where both rows and columns are updated. So the resultant SVD of the incremented matrix will be as given in Eq. (3).

$$\hat{A}_{(m+p) \times (n+q)} = \hat{U}_{(m+q) \times (k+r)} \hat{S}_{(k+r) \times (k+r)} \hat{V}^T_{(k+r) \times (n+p)}, \tag{3}$$
$$\text{where } (k+r) = min(m+q, n+p)$$

Thus, the length of left and right singular vectors will change by q and p respectively. As mentioned earlier, we have adapted the power method [9, 23] for calculating the left and right singular vectors. In the following we focus on the right singular vector (V) since the left singular vector can be calculated from V. It may be noted that the right singular vector diagonalizes the matrix $\hat{A}^T \hat{A}$.

$$\hat{M} = \hat{A}^T \hat{A} \tag{4}$$

The power method is an iterative process in which we initially make some guess about the singular vector or we randomly initialize v^0 (it must not contain all zeros) and then transform it iteratively so that it converges to the actual singular vector. The transformation is given in Eq. (5) below. Given the singular vector for the xth iteration, the power method calculates the singular vector for the $(x + 1)$th iteration as:

$$v_{n\times1}^{x+1} = \hat{M}_{n\times n} \times v_{n\times1}^{x} \tag{5}$$

In Eq. (5) superscripts are number of iterations. After the transformation as above we normalize the singular vector.[2]

$$v_{n\times1} = \frac{v_{n\times1}}{norm(v_{n\times1})} \tag{6}$$

The eigenvalue corresponding to the singular vector can be obtained as follows:

$$d = v' \times \hat{M}_{n\times n} \times v \tag{7}$$

However, as is well known, the method as described above gives the singular vector that corresponds to the largest eigenvalue. For calculating other eigenvalues we use the Deflation Method [23] as described below. Consider a square matrix P whose jth eigenvalue and singular vector pair are given by d_j and u_j i.e.

$$Pu_j = d_j u_j$$

Let the dominant singular vector of P be u_1 and the corresponding eigenvalue be d_1. We can then write

$$(P - d_1 u_1 u_1^T)u_j = Pu_j - d_1 u_1 u_1^T u_j = d_j u_j - d_1 u_1 u_1^T u_j$$

Now if $j = 1$, then

$$(P - d_1 u_1 u_1^T)u_1 = d_1 u_1 - d_1 u_1 (u_1^T u_1) = d_1 u_1 - d_1 u_1 I = 0u_1$$

If $j \neq 1$, then

$$(P - d_1 u_1 u_1^T)u_j = d_1 u_j - d_1 u_1 (u_1^T u_j) = d_1 u_j - d_1 u_1(0) = d_j u_j$$

The above expression implies $(P - d_1 u_1 u_1^T)$ has the same singular vectors and eigenvalues as P except that the largest eigenvalue has been replaced by 0. Thus, the second eigenvalue and singular vector of P can be obtained if we transform P by replacing its largest eigenvalue to zero. The iterative process as defined by Eq. (5) above is again repeated to get the singular vector corresponding to the second largest eigenvalue. The same process can be repeated to obtain all the eigenvalues

---

[2]Right singular vectors of $\hat{A}$ are same as the eigenvectors of $\hat{M}$.

and singular vectors. The process as defined above implies that after obtaining the tth singular vector, we update the $\hat{M}$ as

$$\hat{M}_{t+1} = \hat{M}_t - d_t \times v_t \times v_t^T \tag{8}$$

and then repeat the process for the $(t+1)$th eigenvalue and singular vector. As can be seen, the power method does not guarantee that the singular vectors are orthonormal to each other. To ensure orthonormality we have used Gram-Schmidt orthonormalization only for those vectors which are not orthonormal to others as described next. If $v_1, v_2, v_3, \ldots, v_n$ are some vectors and $u_1, u_2, u_3, \ldots, u_n$ are orthonormal vectors corresponding to $v_1, v_2, v_3, \ldots, v_n$, then Gram-Schmidt orthogonalization process is performed as follows [23]

$$u_k = v_k - \sum_{n=1}^{k-1} proj_{u_j}(v_j) \tag{9}$$

For normalization of the resultant we use

$$u_k = \frac{u_k}{norm(u_k)} \tag{10}$$

While the power method described above is guaranteed to converge even if we initialize the vector randomly, the efficiency of the method described above depends on the initial values of the singular vectors. We have experimented with two different initializations, as described below.

3.1 Method I

Let us assume that the singular vectors and eigenvalues corresponding to the matrix M, have been calculated, where

$$M = A^T A \tag{11}$$

It is required to obtain the eigenvalues and singular vectors corresponding to the incremented matrix, as given by Eq. (4). In Method I, instead of randomly initializing the singular vectors of $\hat{M}$ we use the singular vectors of M. However, we cannot use the singular vectors of M directly since $\hat{M}$ will, in general, have a different size compared to M. As explained earlier, the nature of incrementing is such that the number of rows of $\hat{M}$ can be larger than the number of rows of M (i.e. the length of singular vectors of $\hat{M}$ can be larger than the length of singular vectors of M). In Method I we initialize the singular vectors of $\hat{M}$ to the singular vectors of M and pad with zeros at the end of the vectors. So, the initial values for the singular vectors of $\hat{M}$ are given by

$$\hat{V}_{k \times (n+p)}^T = \left(V_{k \times n} \quad 0_{k \times p} \right) \tag{12}$$

This initialization protects some of the basic properties of the eigenvectors. These are

$$\hat{V}\hat{V}^T = I$$

and

$$norm(\hat{V}) = 1$$

For initialization the remaining eigenvector we simply use the uniform distribution method i.e.

$$\hat{V}^T_{(r \times (n+p))} = \sqrt{\frac{1}{n+p}} \tag{13}$$

This distribution again follows the normalization rule. The pseudo code for this method is given in Algorithm 1.

The above initialization leads to a faster convergence compared to the case where random initialization is employed to initiate the iterative process defined by Eq. (5). Also, the resultant singular vectors are orthonormal and the sum of eigenvalues is equal to the trace of the input matrix.

Algorithm 1 Pseudo Code for Method I

Input: $\hat{M}$, V; where $\hat{M}$ is an $(n+q) \times (n+q)$ square matrix and $\hat{M} = \hat{A}^T \hat{A}$ and V is the old $k \times n$ singular vector matrix

Output: $\hat{V}$, $\hat{S}$; where $\hat{V}$ is $(k+r) \times (n+q)$ eigenvectors of $\hat{M}$ and $\hat{S}$ contains the corresponding $(k+r)(k+r)$ singular values of $\hat{A}$

1: **procedure** METHOD_I($\hat{M}$, V)
2:     **for** i 1 to $(k+r)$ **do**
3:         **if** $i \leq k$ **then**
4:             $v^0 = \begin{pmatrix} V_{i \times n} & 0_{1 \times q} \end{pmatrix}$
5:         **else**
6:             $v^0 = \begin{pmatrix} (\sqrt{\frac{1}{n+q}})_{1 \times (n+q)} \end{pmatrix}$
7:
8:         **while** v has not converged **do**
9:             $v^{(n+1)} = v^n \times \hat{M}$
10:            $v^{(n+1)} = \frac{v^{(n+1)}}{norm(v^{(n+1)})}$
11:            $d = (v^{(n+1)})^T \times \hat{M} \times (v^{(n+1)})$
12:         $\hat{V}[i] \leftarrow v$
13:         $\hat{S}[i,i] \leftarrow \sqrt{d}$
14:         $\hat{M} = \hat{M} - d \times v$
15:     Gram-Schmidt ($\hat{V}$)

3.2 Method II

In the above method we employed only the $(n-1)$th singular vector for calculating the nth singular vector and to remove its residue from the matrix. In method II we employ a different initialization to predict a better approximation for the eigenvalue and singular vector and then remove the residue from the matrix using those predicted values.

We can predict the singular vectors as in Eqs. (12) and (13). However, for predicting the eigenvalues of the incremented matrix, $\hat{M}$, we can use trace property of eigenvalues. If D is the eigenvalue matrix of M then

$$D = S^2 \tag{14}$$

where M is defined by $A^T A$ as given in Eq. (11). The trace property of eigenvalues is

$$trace(M) = Sum(D) \tag{15}$$

If the new matrix is represented by $\hat{M}$ and $\hat{D}$ represents its eigenvalues, then we know the difference in the sums of eigenvalues of both the matrices

$$diff = trace(\hat{M}) - trace(M) = trace(\hat{M}) - Sum(D) \tag{16}$$

This difference in the old and new matrices is going to distribute over all the new eigenvalues of $\hat{M}$.

$$sum(\hat{D}) = sum(D) + diff \tag{17}$$

Now we can distribute this difference uniformly over all eigenvalues i.e.

$$\hat{D}(i) = D(i) + \frac{diff}{k} \tag{18}$$

or we can again make some smart guess about $\hat{D}$ using the same distribution as D. So we can also update $\hat{D}$ as

$$\hat{D}(i) = D(i) + diff * \frac{D(i)}{trace(\hat{M})} \tag{19}$$

Algorithm 2 Pseudo Code for Method II

Input: $\hat{M}$, V, D; where $\hat{M}$ is an $(n + q) \times (n + q)$ square matrix and $\hat{M} = \hat{A}^T \hat{A}$ and V is the old $k \times n$ singular vector matrix and D is old corresponding eigenvalues

Output: $\hat{V}$, $\hat{S}$; where $\hat{V}$ is $(k + r) \times (n + q)$ eigenvectors of $\hat{M}$ and $\hat{S}$ contains the corresponding $(k + r)(k + r)$ singular values of $\hat{A}$

1: **procedure** METHOD_II($\hat{M}$, V, D)
2:    $diff = trace(\hat{M}) - sum(D)$
3:    **for** i 1 to $(k + r)$ **do**
4:      **if** $i \leq k$ **then**
5:        $v^0 = \begin{pmatrix} V_{i \times n} & 0_{1 \times q} \end{pmatrix}$
6:      **else**
7:        $v^0 = \left((\sqrt{\dfrac{1}{n + q}})_{1 \times (n+q)} \right)$
8:      **while** v has not converged **do**
9:        $v^{(n+1)} = v^n \times \hat{M}$
10:       $v^{(n+1)} = v^{(n+1)}/norm(v^{(n+1)})$
11:       $d = (v^{(n+1)})^T \times \hat{M} \times (v^{(n+1)})$
12:       $\hat{D}(i, i) = D(i, i) + diff * \dfrac{D(i)}{trace(M)}$
13:       $\hat{M} = \hat{M} - \hat{D}(i, i) \times v$
14:       $\hat{V}[i] \leftarrow v$
15:       $\hat{D}[i, i] \leftarrow d$
16:       $\hat{S}[i, i] \leftarrow \sqrt{d}$
17:    Gram-Schmidt $(\hat{V})$

Algorithm 3 pseudo code for Gram-Schmidt orthogonalization

Input: $\hat{V}$; where $\hat{V}$ is an $(k + r) \times (n + q)$ eigenvectors of $\hat{M}$

Output: $\hat{V}$; where $\hat{V}$ is an $(k + r) \times (n + q)$ orthogonal eigenvectors of $\hat{M}$

1: **procedure** GRAM- SCHMIDT($\hat{V}$)
2:    $\hat{V}_1 = \dfrac{\hat{V}_1}{norm(\hat{V}_1)}$
3:    **for** $i = 2$ to $(k + r)$ **do**
4:      **for** $j = 1$ to $(i - 1)$ **do**
5:        $\hat{V}_i = \hat{V}_i - (\hat{V}_j.\hat{V}_i)\hat{V}_j$
6:      $\hat{V}_i = \dfrac{\hat{V}_i}{norm(\hat{V}_i)}$

Now we have a guess about eigenvalues (Eq. 19) and singular vectors (Eqs. 12 and 13). Now for Eq. (5) we do not need the actual value and we use these values in place of original eigenvalues and singular vectors. These initial values can now be employed to find the updated singular vectors and eigenvalues by using the power method. The pseudo code for method II is given in Algorithms 2 and 3 gives the pseudo code for Gram-Schmidt orthogonalization.

From above we can get the value of $\hat{V}$. For calculating $\hat{S}$ we take the square root of values of $\hat{D}$. So

$$\hat{S} = sqrt(\hat{D}) \tag{20}$$

The discussions above allow us to calculate the right singular vectors and eigenvalues of the incremented matrix starting from the same quantities of the un-incremented input matrix. The left singular vectors, $\hat{U}$, can be obtained using simple matrix multiplication as given below:

$$\hat{U} = \hat{A}(\hat{V}^T)^{(-1)}\hat{S}^{(-1)} = \hat{A}(\hat{V}^T)^T\hat{S}^{(-1)} = \hat{A}\hat{V}\hat{S}^{(-1)} \tag{21}$$

It may be noted that in method I the singular vectors are calculated in sequential manner i.e. for calculating the $(i + 1)$th singular vector we first calculate the ith singular vector. While in method II we can calculate all the Singular Vectors simultaneously (parallel method).

4 Experiments

We applied the algorithms described above on the task of latent semantic analysis for evaluating document similarity. The experiments were performed on the MEDLINE and TIME[3] corpus. MEDLINE corpus contains 1033 documents and 30 query documents. In order to test our method we have started with 700 documents and built the corresponding term-document matrix. Singular value decomposition is performed on this initial matrix. We then add, one by one, the new documents until we have included all the documents. So the initial matrix is updated 333 times.

TIME corpus contains 423 documents and 83 query documents. We started with a term-document matrix constructed using 200 documents and then added one document at a time until all documents were included. Thus, for the TIME corpus, a total of 223 updates are performed on the initial matrix.

We follow the standard process for calculating document similarity i.e. tokenization, stop-word removal and stemming of words, for each document. At the end we create a term-document matrix for the corpus. We have used the Stanford-NLP library[4] for tokenization, stop-word removal and stemming. We have also compared our results with one of the existing methods given by Jiang et al. [12]. In order to have a meaningful comparison between our results and those of Jiang et al. [12] we ensured that the same term-document matrix was used at each stage for both the algorithms i.e. the same corpus was used at each stage.

5 Results and Discussion

As discussed earlier, the MEDLINE corpus contains 1033 documents and 30 queries. After applying standard preprocessing steps i.e. tokenization, stop-word removal and

---

[3]http://ir.dcs.gla.ac.uk/resources/test_collections.

[4]http://nlp.stanford.edu/software/stanford-corenlp-full-2015-12-09.zip.

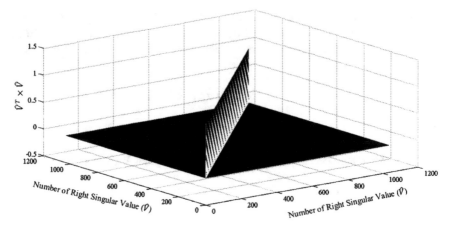

Fig. 1 Orthonormality property of right singular vector ($\hat{V}^T \times \hat{V}$) for method I

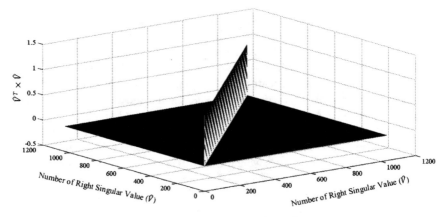

Fig. 2 Orthonormality property of right singular vector ($\hat{V}^T \times \hat{V}$) for method II

stemming we consider only those terms which occur in more than one document. This leads to 4855 distinct terms leading to a term-document matrix with a size of 4855×1033. We have taken 700 documents initially (with 4519 distinct terms), leading to an initial input matrix of size 4519×700. Standard batch mode SVD was applied on this matrix. The incremental methods, described above, were then applied by incrementing this initial matrix with new documents. The documents were added one by one leading to addition of new rows and columns till the full size of 4855×1033 was achieved. Thus, a total of 333 increments of the initial matrix were performed.

After updating all three matrices (i.e. U, S and V) for 333 times we found that the final matrices ($\hat{U}$, $\hat{S}$ and $\hat{V}$) conform to their properties of orthonormality, trace etc. The results of the first property, that is the orthogonality property of left and right singular vectors is shown in Fig. 1 (for method I) and Fig. 2 (for method II). In both these we show the values of $v_i^T v_j$ for all values of i and j.

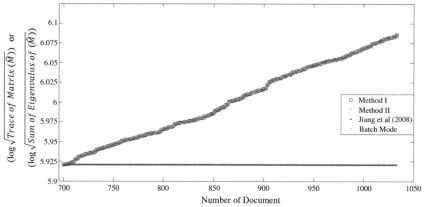

(a) A plot between trace of matrix and sum of square of singular values of matrix

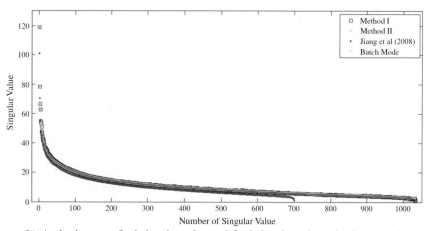

(b) A plot between final singular value and final singular value using batch mode

Fig. 3 Eigenvalue properties with MEDLINE corpus

The other property which our methods satisfy is that the sum of eigenvalues (sum of square of singular values) is equal to the trace of the matrix. Figure 3a shows the results of both methods of initialization along with trace of matrix $\hat{M}$. After each step our results are quite similar to the actual trace of matrix. It also contains the results obtained using one of the earlier methods [12].

Figure 3b shows the final singular values calculated by both methods along with the singular value that would be obtained by using any batch mode SVD technique. Results of both the methods are quite similar and they match closely with the result obtained using batch mode SVD. These values remain quite close even for small singular values. Moreover, the results obtained using both methods do not deviate from the exact results even after repeated increments. In other words, there is no degradation of the results even after repeated increments and we do not have to run the original SVD after a few increments.

The convergence rate of the initialization methods is substantially better as compared to actual convergence of power method with random initialization. The estimated convergence rate, with random initialization, is proportional to $|\lambda_{(i+1)}/\lambda_i|$. The above expression is obtained assuming that our singular values are accurate up to 4 significant figures. Using the above expression we can estimate the number of iterations required for convergence of singular value as

$$Number\ of\ Iterations\ (noi) = \frac{\log_{10}(10^{-4})}{\log_{10}(|\frac{\lambda_{i+1}}{\lambda_i}|)}$$

where noi is number of iterations required to calculate ith eigenvalue, λ_i and λ_{i+1} are the ith and $(i + 1)$th eigenvalues respectively. If we know the values of λ_i and λ_{i+1} then we can calculate the value of noi if we want results with an accuracy

Table 1 comparison of convergence rate for power method and given methods

Singular value (S)	Eigen value (S^2)	Convergence rate ($|\frac{\lambda_2}{\lambda_1}|$)	$-\log_{10}(|\frac{\lambda_2}{\lambda_1}|)$	noi	Actual number of iterations
118.965	14152.671	0.4345	0.362	11.05	5 (Method I)
118.9739	14154.789	0.4345	0.362	11.05	9 (Method II)
78.4204	6149.7591	0.7205	0.1424	28.1	17 (Method I)
78.422	6150.0101	0.7211	0.142	28.16	23 (Method II)
66.5658	4431.0057	0.8829	0.0541	73.96	13 (Method I)
66.5923	4434.5344	0.8825	0.0543	73.66	100 (Method II)
62.5474	3912.1773	0.7626	0.1177	33.99	20 (Method I)
62.5563	3913.2907	0.7632	0.1173	34.09	27 (Method II)
54.6212	2983.4755	0.9549	0.0201	199.53	42 (Method I)
54.6509	2986.7209	0.9534	0.0207	193.02	100 (Method II)
53.375	2848.8906	0.9133	0.0394	101.58	25 (Method I)
53.3624	2847.5457	0.9134	0.0393	101.67	100 (Method II)
51.0093	2601.9487	0.8784	0.0563	71.01	28 (Method I)
50.9993	2600.9286	0.883	0.054	74.05	61 (Method II)
47.8062	2285.4328	0.9415	0.0262	152.89	22 (Method I)
47.9244	2296.7481	0.938	0.0278	143.88	100 (Method II)
46.3877	2151.8187	0.9145	0.0388	103.07	27 (Method I)
46.4148	2154.3337	0.914	0.0391	102.42	100 (Method II)
44.3608	1967.8806				(Method I)
44.374	1969.0519				(Method II)

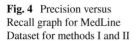

Fig. 4 Precision versus
Recall graph for MedLine
Dataset for methods I and II

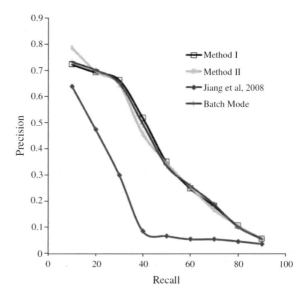

of 4 significant figures. The estimated value of noi, together with the actual values
observed using the two methods of initialization proposed in this work, are given in
Table 1 for the first nine eigenvalues.

From Table 1 it is clear that method I requires less number of iterations compared
to Method II. However, both of them require significantly less number of iterations
as compared to the estimated number of iterations (noi).

The singular vectors obtained was applied to the task of Latent Semantic Indexing
(LSI) [8] for finding document similarity in the MEDLINE corpus. The Precision-
Recall curve is given in Fig. 4 for both the methods. The results obtained by us
are compared with those obtained by using a Batch-mode SVD as well as the iLSI
method given by [12]. Both the proposed methods give results that are in close
agreement with those obtained using the batch mode SVD. On the other hand, the
results obtained using iLSI show a steady divergence from the exact result. These
results clearly demonstrate that the proposed methods can outperform the existing
methods when we expect a steady arrival of new data which is the hallmark of any
online system.

We check the accuracy of our system on one more corpus i.e. TIME corpus.
This corpus contains 423 documents with 83 query documents. We get 8887 distinct
terms after applying the preprocessing steps (tokenization, stop-word removal and
stemming) and considering only those terms which occurs in more than one document. This leads to a final term document matrix of size 423×8887. We started with
200 documents initially (with 7511 terms) and performed batch mode SVD on this
matrix. Incremental methods described above were then applied in the same fashion
as for the MEDLINE corpus. The results of our methods along with batch mode
method are given in Fig. 5.

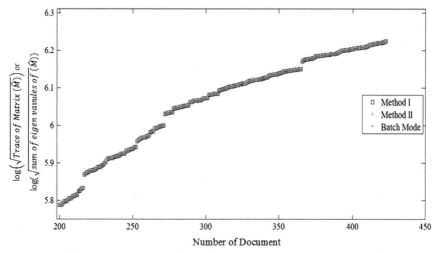

(a) A plot between trace of matrix and sum of square of singular values of matrix

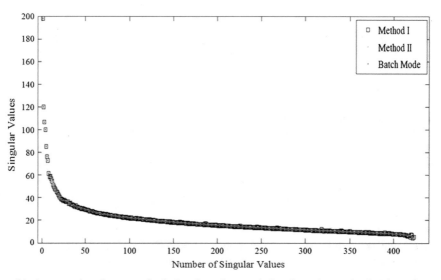

(b) A comparison between final singular values and singular values using batch mode

Fig. 5 Eigenvalue properties using TIME corpus

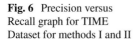

Fig. 6 Precision versus
Recall graph for TIME
Dataset for methods I and II

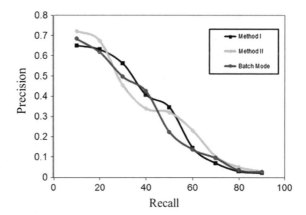

In Fig. 5a plot between trace and sum of square of singular values is presented for
both the methods and batch mode method. The graph of both methods overlaps with
batch mode method for all 223 increments. Similarly Fig. 5b shows a plot of final
singular values by both method and batch mode method. Here again final singular
values of both methods are same as batch mode method. So we can say ISVD method
I and II are consistent over different corpus.

We again apply the results of our methods along with the batch mode method
for measuring document similarity for TIME corpus. The precision recall curves are
given in Fig. 6. Method I and II produce the same results as the batch mode SVD.

6 Conclusion

In this work we have proposed an adaptation of the power method to perform Incremental Singular Vector Decomposition. We have examined two different methods of
initialization. Both methods give similar results and these are quite close to results
obtained using exact methods. Experiments performed using the MEDLINE and
TIME corpus show that the results obtained by us do not deviate from the exact
results even after more than hundreds of increments in the original input matrix. The
convergence rates of both the proposed methods are significantly less than the estimated convergence rates. Thus, we expect that the proposed method will be useful
for several online problems that have a steady arrival of new data which need to be
modeled in real time.

Acknowledgements The authors gratefully acknowledge the infrastructural support provided by
Indian Institute of Information Technology, Allahabad (IIIT-A). One of the authors (SG) also
acknowledges the financial support from IIIT-A.

References

1. Baker, C.G., Gallivan, K.A., Dooren, P.V.: Low-rank incremental methods for computing dominant singular subspaces. Linear Algebra Appl. **436**, 2866–2888 (2012)
2. Balsubramani, A., Dasgupta, S., Freund, Y.: The fast convergence of incremental PCA. In: Advances in Neural Information Processing Systems (2013)
3. Brand, M.: Incremental singular value decomposition of uncertain data with missing values. In: Proceedings of the 7th European Conference on Computer Vision-Part I, London, UK, pp. 707–720 (2002)
4. Brand, M.: Fast low-rank modifications of the thin singular value decomposition. Linear Algebra Appl. **415**, 20–30 (2006)
5. Bhushan, A., Sharker, M.H., Karimi, H.A.: Incremental principal component analysis based outlier detection methods for spatiotemporal data streams. ISPRS Ann. Photogramm. Remote Sens. Spat. Inf. Sci. **2**(4), 67 (2015)
6. Chandrasekaran, S., Manjunath, B.S., Wang, Y.F., Winkeler, J., Zhang, H.: An Eigenspace update algorithm for image analysis. In: CVGIP: Graphical Model and Image Processing, vol. 59, pp. 321–332 (1997)
7. Davies, P.I., Smith, M.I.: Updating the singular value decomposition. J. Comput. Appl. Math. **170**(1), 145–167 (2004)
8. Deerwester, S., Dumais, S.T., Furnas, G.W., Landauer, T.K., Harshman, R.: Indexing by latent semantic analysis. J. Am. Soc. Inf. Sci. **41**, 391–407 (1990)
9. Golub, G.H., Van Loan, C.F.: Matrix Computations, 3rd edn. Johns Hopkins University Press, Baltimore (MD) (1996)
10. Hall, P.M., David Marshall, A., Martin, R.R.: Incremental eigenanalysis for classification. In: BMVC, vol. 98 (1998)
11. Hao, S., Xu, Y., Ke, D., Su, K., Peng, H.: SCESS: a WFSA-based automated simplified chinese essay scoring system with incremental latent semantic analysis. Nat. Lang. Eng. **22**, 291–319 (2016)
12. Jiang, H.Y., Nguyen, T.N., Chen, I.X., Jaygarl, H., Chang, C.K.: Incremental latent semantic indexing for automatic traceability link evolution management. In: 23rd IEEE/ACM International Conference on Automated Software Engineering, 2008. ASE 2008, pp. 59–68 (2008)
13. Kaufman, L.: Methods for updating the singular value decomposition. DAIMI Report Series 3 (1974)
14. King, J.D., Li, Y.: Web based collection selection using singular value decomposition. In: Web Intelligence, pp. 104–110 (2003)
15. Kwok, J.T., Zhao, H.: Incremental eigen decomposition. Matrix **100**(C1), C2 (2003)
16. Levy, A., Lindenbaum, M.: Sequential Karhunen-Loeve Basis Extraction and Its Application to Images
17. Oyama, T., et al.: Incremental learning method of simple-PCA. In: International Conference on Knowledge-Based and Intelligent Information and Engineering Systems. Springer, Berlin, Heidelberg (2008)
18. Ozawa, S., Pang, S., Kasabov, N.: A modified incremental Principal Component Analysis for on-line learning of feature space and classifier. In: Pacific Rim International Conference on Artificial Intelligence. Springer, Berlin, Heidelberg (2004)
19. Press, W.H., Teukolsky, S.A., Vetterling, W.T., Flannery, B.P.: Numerical Recipes in C: The Art of Scientific Computing, 2nd edn. Cambridge University Press, New York, NY, USA (1992)
20. Rao, S., Medeiros, H., Kak, A.: Comparing incremental latent semantic analysis algorithms for efficient retrieval from software libraries for bug localization. SIGSOFT Softw. Eng. Notes. **40**, 1–8 (2015)
21. Rehurek, R.: Subspace tracking for latent semantic analysis. In: Clough, P.D., Foley, C., Gurrin, C., Jones, G.J.F., Kraaij, W., Lee, H., Murdock, V. (eds) ECIR, pp. 289–300 (2011)
22. Ross, D.A., Lim, J., Lin, R.S., Yang, M.H.: Incremental learning for robust visual tracking. Int. J. Comput. Vis. **77**(1–3), 125–141 (2008)

23. Saad, Y.: Numerical methods for large eigenvalue problems. In: SIAM (2011)
24. Sabes, P.N.: Linear Algebraic Equations, SVD, and the Pseudo-Inverse. http://keck.ucsf.edu/
25. Sarwar, B., et al.: Incremental singular value decomposition algorithms for highly scalable recommender systems. In: Fifth International Conference on Computer and Information Science (2002)
26. Stewart, G.W.: An updating algorithm for subspace tracking. IEEE Trans. Signal Process. **40**(6), 1535–1541 (1992)
27. Stewart, M., Van Dooren, P.: Updating a generalized URV decomposition. SIAM J. Matrix Anal. Appl. **22**(2), 479–500 (2000)
28. Yang, J., et al.: A fast incremental algorithm for low rank approximations of matrices and its applications in facial images. J. Univ. Sci. Technol. China **39**(9), 970–979 (2009)
29. Zhang, M., Hao, S., Xu, Y., Ke, D., Peng, H.: Automated essay scoring using incremental latent semantic analysis. J. Softw. **9** (2014)
30. Zha, H., Simon, H.D.: On updating problems in latent semantic indexing. SIAM J. Sci. Comput. **21**, 782–791 (1999)
31. Zhao, H., Yuen, P.C., Kwok, J.T.: A novel incremental principal component analysis and its application for face recognition. IEEE Trans. Syst. Man Cybern. Part B (Cybern.) **36**(4), 873–886 (2006)

Personalized Landmark Recommendation for Language-Specific Users by Open Data Mining

Siya Bao, Masao Yanagisawa and Nozomu Togawa

Abstract This paper proposes a personalized landmark recommendation algorithm aiming at exploring new sights into the determinants of landmark satisfaction prediction. We gather 1,219,048 user-generated comments in Tokyo, Shanghai and New York from four travel websites. We find that users have diverse satisfaction on landmarks those findings, we propose an effective algorithm for personalize landmark satisfaction prediction. Our algorithm provides the top-6 landmarks with the highest satisfaction to users for a one-day trip plan our proposed algorithm has better performances than previous studies from the viewpoints of landmark recommendation and landmark satisfaction prediction.

Keywords User-generated comment · Landmark satisfaction prediction Landmark recommendation

1 Introduction

User-generated comments represent users' personal travel experiences and users' tourism decisions are strongly influenced by comment contents in many aspects [1].

Several studies focus on predicting users' satisfaction on landmarks through social media comments [2], some of which try to discover the connectedness between users' travel behaviors and backgrounds [3]. Those studies take into consideration new ideas for efficiently predicting users' satisfaction on landmarks but their accuracy is often limited to the sample size (mostly, the number of samples $N \leq 600$) and the usage of a single data source. Moreover, deviance in language discrepancies is not considered

S. Bao (✉) · M. Yanagisawa · N. Togawa
Department of Computer Science and Communications Engineering,
Waseda University, Tokyo, Japan
e-mail: siya.bao@togawa.cs.waseda.ac.jp

M. Yanagisawa
e-mail: myanagi@waseda.jp

N. Togawa
e-mail: togawa@togawa.cs.waseda.ac.jp

© Springer Nature Switzerland AG 2019
R. Lee (ed.), *Computer and Information Science*, Studies in Computational
Intelligence 791, https://doi.org/10.1007/978-3-319-98693-7_8

Table 1 The statistics of user comment among Ctrip, Jaran and TripAdvisor in Shanghai

	User comment	Chinese	English	Japanese	Other-languages
Ctrip	66,282	>99%	<1%	<1%	<1%
4travel	2598	<1%	<1%	>99%	<1%
TripAdvisor	68,563	7357 (10.73%)	29,316 (42.76%)	14,137 (20.62%)	17,753 (25.89%)
Total	137,443	73,639 (53.58%)	29,316 (21.33%)	16,735 (12.18%)	17,753 (12.92%)

though users from different backgrounds have different satisfaction rating behaviors [4].

In this paper, we utilize large datasets using heterogeneous open data sources and divide users into three languages (Chinese, English and Japanese) groups. We firstly collect 1,219,048 user-generated comments from the travel websites of Ctrip [5], Jaran [6], 4travel [7] and TripAdvisor [8] for 194 landmarks in Tokyo, Japan, 189 landmarks in Shanghai, China and 196 landmarks in New York, USA. We analyze users' average satisfaction on landmarks and landmark coverage differences between the travel websites in three cities. Then we extract 1,046,395 user comments from 1,219,048 user comments and divide them into three languages groups. Users' language-specific satisfaction and favorable landmark types are examined. Finally, an algorithm is proposed to predict the users' satisfaction over each landmark according to their preferences on landmark types, languages and travel websites.

Our contributions are highlighted as follows:

- We analyze data-source-specific and language-specific landmark satisfaction differences with 1,219,048 user comments through four travel websites.
- We analyze pairwise landmark type relationships in order to correct error types in the existing travel websites.
- We propose a personalized landmark recommendation algorithm based on landmark satisfaction prediction. Our algorithm can recommend landmarks that fit the user's preferences with an accuracy over 82% and successfully predicts the user's satisfaction on landmark with the error rate lower than 7.5%, which outperforms the previous studies.

2 Data Source and Data-Collect Process

We collect user comments from four leading travel websites in Ctrip (China) [5], Jaran (Japan) [6], 4travel (Japan) [7], and TripAdvisor (United States) [8], with the aim to deal with the small sample size issue in the previous studies as discussed in

Sect. 1. A data-collection program was developed in R, which took approximately 20 days to crawl all the data.

All user comments collected from each website was before February 1st, 2018, and Tokyo, Shanghai and New York are famous travel destinations. Particularly, the landmark which is labeled as "region", for example, "Shibuya District", is not considered as it does not have a specific type or location. Let N_L be the number of total landmarks in each website. Then we have for Tokyo, $N_L = 835$ in Ctrip, $N_L = 1140$ in Jaran, and $N_L = 1336$ in TripAdvisor; for Shanghai, $N_L = 4420$ in Ctrip, $N_L = 1497$ in 4travel, and $N_L = 1680$ in TripAdvisor, and for New York, $N_L = 631$ in Ctrip, $N_L = 227$ in 4travel, and $N_L = 4450$ in TripAdvisor. We extracted top-k ranked landmarks (we set $k = 100$), each of which has at least five user comments from each website.[1] For each landmark in a travel website, we collected each user's satisfaction on the landmark, comment content and the time of writing. User's satisfaction on each landmark is five-scale rated from 1 (very dislike) to 5 (very like). We also collected the rank of the landmark at each travel website.[2] As a result, 1,219,048 user comments were collected for further research (see Table 1).

After data collection, we conducted language-detection through R. Three representative characters, including Chinese (de), English (is) and Japanese (no), were utilized to distinguish a specific language.

As limited space, an example of the statistics of user comments from three travel websites in Shanghai is shown in Table 1. As a result, the statistics show that Ctrip mainly includes Chinese users as the ratio of Chinese comments is always over 99%, and thus we consider Ctrip's comments all as Chinese comments. Similarly, we consider Jaran's comments all as Japanese comments. In cases of Tokyo and New York, we also consider Ctrip's comments all as Chinese comments, and Jaran's and 4travel's all as Japanese comments. In addition, other languages in TripAdvisor provides a ratio no more than 13.89% of the total comment set. For this reason, we only concentrate on Chinese, English and Japanese for language-specific analysis in Sect. 4.

3 Data-Source-Specific Analysis

3.1 Comparison of Landmark Coverage

We investigate the coverage ratio of top-k ranked landmarks between the websites in three cities (see Fig. 1). "Coverage" is the ratio of landmarks in one website's top-k ranked landmarks that also occur in the other websites' top-k ranked landmarks. The

---

[1]Since Jaran is the website for Japanese sightseeing, we extract user comments for Tokyo using it but it does not contain landmarks in other countries. Instead, we use 4travel.jp to extract user comments including Japanese for Shanghai and New York.

[2]Note that, how to rank landmarks at every travel website is not open. It can be just decided based on the user satisfaction and the number of user comments on each website.

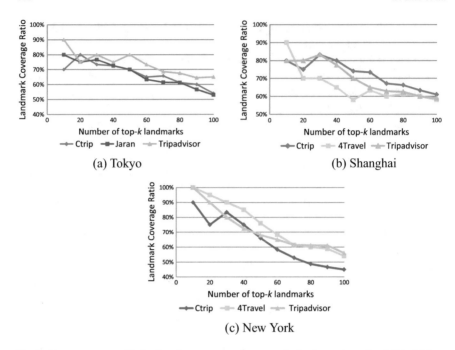

Fig. 1 Data-source-specific landmark coverage ratio among Ctrip, Jaran, 4travel and TripAdvisor

lower the coverage ratio is, the higher the landmark uniqueness of each website is. Figure 1 shows that the coverages of landmarks dramatically differ in the different websites in all city cases. For example, in the case of New York, when $k = 10$, the coverage ratios of all the three websites are more than 90%. When $k \geq 50$, the coverage ratios of all the three websites degrade to 76% or less.

Filtering is one of the most convenient methods in the landmark recommendation [9] but it usually assumes that the coverages of landmarks in all data sources always stay 100%. Oppositely, our results suggest that the coverages of landmarks remarkably vary by the travel websites. Therefore, our findings highlight the importance of collaborating heterogeneous data sources to resolve the problem of usage of a single data source as we discuss in Sect. 1.

3.2 Comparison of Average Satisfaction

The data-source-specific average satisfaction AS_k^{web} of the top-k landmarks in a travel website *web* is defined as follows:

$$AS_k^{web} = \frac{\sum_{i=1}^{k} ds(l_i)^{web}}{k} \tag{1}$$

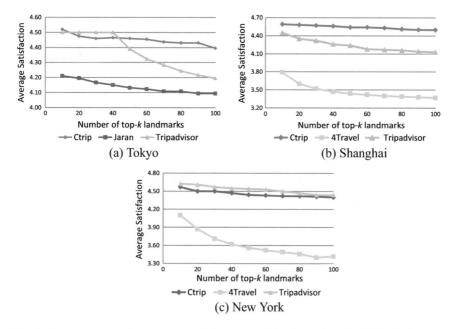

Fig. 2 Data-source-specific average satisfaction among Ctrip, Jaran, 4travel and TripAdvisor

where $ds(l_i)^{web}$ is the data-source-specific satisfaction and it is the average of all users' satisfaction for the i-th ranked landmark l in a particular travel website. Then we compare the differences between average satisfaction for the top-k ranked landmarks in the four websites (see in Fig. 2).

In Fig. 2, Jaran always has the lowest average satisfaction in the case of Tokyo, and 4travel has the lowest average satisfaction in the cases of Shanghai and New York. On the other hand, Ctrip and TripAdvisor has higher average satisfaction than Jaran or 4travel. This might be explained as the differences between various travel website user groups. In other words, Ctrip and TripAdvisor users tend to have a higher satisfaction rate on landmarks.

3.3 Comparison of Type Information

A landmark type represents the characteristics of the landmark. However, every travel website has its own way to describe types. Thus we have re-arranged all landmarks' types into eight types of *Art, Entertainment, Food and drink, History, Nature, Night life, Shopping*, and *Sport*. Let *LT* be a set of these eight landmark types. It is worth mentioning that, a landmark can have more than one types, for example, "Statue of Liberty" has two landmark types of *Art* and *History*.

Table 2 Landmark type comparison with Ctrip, Jaran, 4travel and TripAdvisor

	k	Total landmark type	Average landmark type	Improvement rate (%)
Tokyo				
Ctrip	100	146	1.46	+23.97
Jaran	100	178	1.78	+1.69
TripAdvisor	100	158	1.58	+14.56
Ours	194	351	1.81	–
Shanghai				
Ctrip	100	178	1.78	+2.25
4travel	100	161	1.61	+13.05
TripAdvisor	100	152	1.52	+19.74
Ours	189	344	1.82	–
New York				
Ctrip	100	184	1.84	+13.13
4travel	100	163	1.63	+27.71
TripAdvisor	100	205	2.05	+1.54
Ours	196	408	2.08	–

We combine the landmarks and their types of the top-100 ranked landmarks in each website and the results are shown in Table 2. In an instance of landmark combination, we have 100 landmarks in each website in Tokyo and obtain totally 194 different landmarks by discarding the redundant ones. In detail, "Yu Garden" in Shanghai is labeled as *Art* and *History* in Ctrip, *Nature* in TripAdvisor and *Art* and *Nature* in 4travel. Then we combine all the types in the three websites and discard the redundant parts. The refined types for "Yu Garden" are *Art*, *History* and *Nature*.

It is interesting that a travel website provides more type information for its local or domestic landmarks compared with foreign landmarks. This can be explained by the travel website companies are more familiar with the local landmarks, and may describe the landmarks with more details.

Our average type number is of a maximum improvement rate of 27.71% compared with the original four travel websites (see Table 2). The results indicate that it is necessary to collaborate information among different data sources to enrich landmark type information.

4 Language-Specific Analysis

4.1 Comparison of Average Satisfaction

We analyze the user average satisfaction based on language discrepancies in three cities using the user comments in TripAdvisor. In total, 92523 user comments are used for the case of Tokyo; 50810 user comments are used for the case of Shanghai and 674219 user comments are used for the case of New York.

The language-specific average satisfaction for Chinese is calculated as the average satisfaction of all Chinese users' satisfaction on the top-k ranked landmarks in each city as follows. The average satisfaction for Japanese and English is calculated similarly.

$$AS_k^{lang,TripAdvisor} = \frac{\sum_{i=1}^{k} ls^{lang,TripAdvisor}(l_i)}{k} \tag{2}$$

where $ls^{lang,TripAdvisor}(l_i)$ is the average of all *lang* users' satisfaction for the i-th ranked landmark l_i in TripAdvisor, where *lang* is either of Chinese, English or Japanese.

Figure 3 portrays the results of language-specific average satisfaction in three cities. Chinese and English users average satisfaction is relatively similar and is always higher than Japanese users' average satisfaction. Even though Japan shares a great cultural similarity with China, Japanese users' average satisfaction is significantly deviating from the Chinese groups.

To sum up, the results say that users from different cultural backgrounds value the landmark's average satisfaction differently and the conclusion stays the same no matter which city's data we use. Therefore, it is crucial to examine the users' language backgrounds into consideration for accurate satisfaction prediction.

4.2 Comparison of Overall Type Preference

We assume that all users have similar preferences on landmark types no matter which language they use and we demonstrate our assumption as follows:

We calculate frequencies and ranks of the eight landmark types for the three language groups using 1,038,087 user comments for each city. In order to visualize the similarities between the three language groups, we calculate the cosine similarity *cosim* of the type preferences in two ways:

We first define a vector

$$\mathbf{t_r}(\mathbf{lang}) = (r_{Art}, r_{Entertainment}, \ldots, r_{Sport}) \tag{3}$$

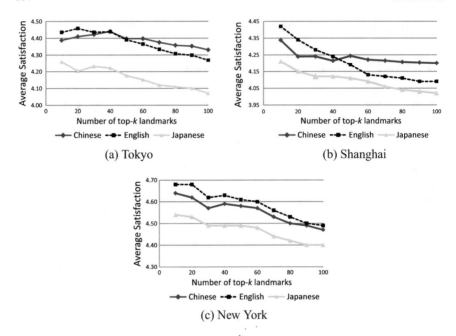

(a) Tokyo   (b) Shanghai

(c) New York

Fig. 3 Language-specific average satisfaction among Chinese, English and Japanese

for a language *lang*, where $\mathbf{r_t}$ is defined that, if the landmark type t is the i-th rank in the language *lang*, then r_t is $(1/i)$ in this language in each city. For example, in the case of $\mathbf{t_r}$(**Chinese**) in Shanghai, *Art* is the second place then $r_{Art} = 1/2 = 0.5$ and *Sport* is the last place with $r_{Sport} = 1/8 = 0.125$.

In the same way, we define the other vector as follows:

$$\mathbf{t_f}(\mathbf{lang}) = (f_{Art}, f_{Entertainment}, \ldots, f_{Sport}) \tag{4}$$

for a language *lang*, where f_t is the frequency of the landmark type t in the language *lang* in each city. For example, in the case of $\mathbf{t_f}$(**Chinese**) in Shanghai, f_{Art} and f_{Sport} become 0.26 and 0.00, respectively.

Then Eq. (5) shows the $cosim_r$ value weighted by the type rank and Eq. (6) shows the $cosim_f$ value weighted by the type frequency:

$$cosim_r(lang_i, lang_j) = \frac{<\mathbf{t_r}(\mathbf{lang_i}) \cdot \mathbf{t_r}(\mathbf{lang_j})>}{\|\mathbf{t_r}(\mathbf{lang_i})\| \cdot \|\mathbf{t_r}(\mathbf{lang_j})\|} \tag{5}$$

$$cosim_f(lang_i, lang_j) = \frac{<\mathbf{t_f}(\mathbf{lang_i}) \cdot \mathbf{t_f}(\mathbf{lang_j})>}{\|\mathbf{t_f}(\mathbf{lang_i})\| \cdot \|\mathbf{t_f}(\mathbf{lang_j})\|} \tag{6}$$

where $<\mathbf{v_1} \cdot \mathbf{v_2}>$ shows the inner product of two vectors $\mathbf{v_1}$ and $\mathbf{v_2}$ and $\|\cdot\|$ shows the L^2 norm.

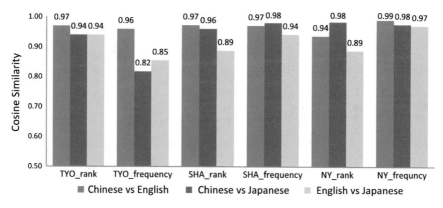

Fig. 4 Cosine similarity between three language groups

The results of *cosim_r* and *cosim_f* are shown in Fig. 4. It can be seen that the $cosim_r$ values or the $cosim_f$ values between the three language groups are of high similarities over 0.82. The results indicate that the users' overall preferences of landmark types do not have a direct association with the language that they use. This confirms our assumption at the beginning of this subsection.

Thus, we concentrate on the users' personalized preferences on landmark types, rather than taking the users' overall preferences into account.

5 Landmark Database Establishment

We build a landmark database of three cities. After combining types and eliminating redundant landmarks, 194 landmarks are kept for Tokyo, 189 landmarks are kept for Shanghai and 196 landmarks are kept for New York.

Unfortunately, error types still exist in some cases. For example, in the case of Jimbocho Bookstore Area (a bookstore street), it is labeled as *Art* + *History* + *Shopping* in one website, where *Shopping* seems not proper. Thus, we analyze the relationship between type pairs. Figure 5 presents the top-3 strong relations between the eight types. The size of the circle of each type shows the frequency at which it occurs in the database and the thickness of an arrow link between a pair of types shows how strong the relationship is.

Then we conduct the pairwise comparison between types for each landmark. If one pair of types is not included in the top-3 relations, then it refers that the relation between the pair is not strong enough and it will be labeled as a potential error pair. Error pairs are manually re-checked. For example, *History* does not has a strong relation with *Shopping* and we discard the type of *Shopping* for Jimbocho Bookstore Area.

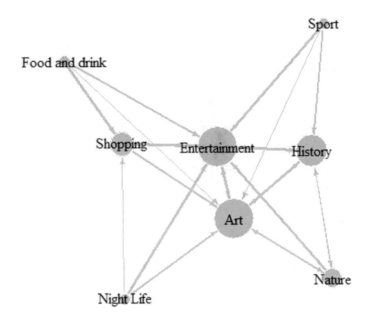

Fig. 5 Representation of pairwise type relations

Our finalized database contains 579 landmarks and 1103 types (see Table 2). Each landmark l in the finalized database include:

- a language-specific satisfaction rating $ls^{lang}(l)$ for each language $lang$, which is defined by the average satisfaction over all the comments to l in a specific language.
- a data-source-specific satisfaction satisfaction $ds^{web}(l)$ for every travel website web, which is defined by the average satisfaction over all the comments to l in a specific website.

6 Satisfaction Prediction

In order to predict user's satisfaction on every landmark, we propose a mathematical model to simulate the relation between user satisfaction and three variables, which are user's preferences on landmark types, user's language(s) and commonly visited travel website(s). We consider a linear relation, which is used in many related studies [10].

We introduce $S_{u,l}$ as the prediction satisfaction on a landmark l of a user u. $S_{u,l}^{type}$ is the satisfaction on l depending on u's type preferences. Likewise, $S_{u,l}^{lang}$ and $S_{u,l}^{web}$ are the the satisfaction on l depending on u's commonly used language(s) and travel website(s), respectively. $S_{u,l}$ is as follows:

$$S_{u,l} = \alpha \times S_{u,l}^{type} + \beta \times S_{u,l}^{lang} + (1 - \alpha - \beta) \times S_{u,l}^{web} + \theta \qquad (7)$$

where α and β are the two constants weighting the significance of the three variables $(0 \le (\alpha + \beta) \le 1)$. In an instance, if $\alpha = 1$, $S_{u,l}$ is only affected by the type preferences. We have both α and β equal to $1/3$ for the following analysis, which we assume that the three variables to be of equivalent importance. The calculation of the three variables is introduced as follows:

For $S_{u,l}^{type}$, firstly, a user is required to rate his/her preference $w(t)$ on each landmark type t with a five-scale rating: very dislike (1), dislike (2), fair (3), like (4), and very like (5). Let $LT(l)$ be a set of types that the landmark l contains and then $S_{u,l}^{type}$ is computed as follows:

$$S_{u,l}^{type} = \frac{\sum_{t \in LT(l)} w(t)}{|LT(l)|} \qquad (8)$$

where $|LT(l)|$ is how many landmark types that $LT(l)$ has.

For $S_{u,l}^{lang}$, $ls^{lang}(l)$ is a language-specific satisfaction of a landmark l for a specific language $lang$ and $LS(l)$ is a set of language-specific satisfactions of all the languages that l has. If the user's language is $lang$ and l has the corresponding satisfaction $ls^{lang}(l)$, then $S_{u,l}^{lang}$ is equal to $ls(l)^{lang}$. Otherwise, $S_{u,l}^{lang}$ is the average of its other languages satisfaction as below:

$$S_{u,l}^{lang} = \begin{cases} ls^{lang}(l) & (\text{if } LS(l) \text{ includes } lang) \\ \frac{\sum_{\text{all languages}} ls^{lang}(l)}{|LS(l)|} & (\text{otherwise}) \end{cases} \qquad (9)$$

where $|LS(l)|$ represents how many language-specific satisfaction that $LS(l)$ has.

For $S_{u,l}^{web}$, $ds^{web}(l)$ is a data-source-specific satisfaction in a specific travel website web for l and $DS(l)$ is a set of data-source-specific satisfactions of all the websites that l has. If the user's preferred website is web, and l has the corresponding satisfaction $ds^{web}(l)$, then $S_{u,l}^{web}$ is equal to $ds^{web}(l)$. Otherwise, $S_{u,l}^{web}$ is the average of its other travel website satisfaction as below:

$$S_{u,l}^{web} = \begin{cases} ds^{web}(l) & (\text{if } DS(l) \text{ includes } web) \\ \frac{\sum_{\text{all websites}} ds^{web}(l)}{|DS(l)|} & (\text{otherwise}) \end{cases} \qquad (10)$$

where $|DS(l)|$ shows how many travel website satisfaction that $DS(l)$ includes.

In addition, we set a bonus constant θ as an error parameter [10]. Let $\theta = (n(l) - 1) + 0.1$, where $n(l)$ is the number of times the landmark l occurs in the top-100 ranked landmarks in a city among the three travel websites.

7   Evaluation

In this paper, the experimental areas are set to be Tokyo, Shanghai and New York and the 194, 196 and 189 landmarks in our database are, respectively, used for landmark recommendation for each city. The algorithm is coded in R.

We test our algorithm with 12 different user profiles. We have interviewed 12 users, aging from 20 to 60, 4 males and 8 females. They are 6 Japanese users, 4 Chinese users and 2 English users. They are also 4 Ctrip users, 4 Jaran users, 1 Jaran and 4travel user and 3 TripAdvisor users. 12 users have visited at least three landmarks in Tokyo, Shanghai and 10 users have visited at least three landmarks in New York.

7.1   Landmark Type Recommendation Precision

In this subsection, landmark type recommendation precision of our algorithm is analyzed. Users were required to fill in a questionnaire about the preferences with the five-scale rating on the landmark types, also their commonly used languages, and commonly used travel websites were recorded.

The proposed algorithm first calculates the $S_{u,l}$ values for each landmark l in the database by the user's profile. Duration time on a landmark usually takes around 1–2 h, we assume that a user visits at most 6 landmarks in a one-day trip.

Our algorithm provides each user with the top-6 landmarks with the highest prediction satisfaction $S_{u,l}$. Three other algorithms are used for comparisons. Random algorithm randomly recommends six landmarks through the top-100 ranked landmarks in TripAdvisor, Popular-first algorithm recommends the top-6 ranked landmarks in TripAdvisor and the algorithm in [2] recommends six landmarks that match the user's type preferences. The six recommended landmarks are recorded as l_i^{rec} ($i = 1, 2, \ldots, 6$).

Next, we compare the precision of the six recommended landmarks of our proposed algorithm with that of the other three comparison algorithms. We consider that the landmark type t fits the user's preferences if the user has rated three points or more to t. Then the number of true positives (TP) is defined by the number of the recommended landmarks' types that successfully fits the user's preferences. We consider that the landmark type t fails to fit the user's preferences if the user has rated two points or less to t. Then the number of false positives (FP) is defined by the number of the recommended landmarks' types that fails to fit the user's preferences.

For example, in the case of Tokyo, assume that the ratings of the user u_1 to the landmark types of *Art* and *Nature* are four and five points, respectively. Assume also that our algorithm recommends for the user u_1 the first landmark $l_1^{rec} = Sensoji$, of which types are labeled by *Art* and *Nature* in our database. Then TP for *Sensoji* is two and its FP is zero, since the two types *Art* and *Nature* fit the user's preferences. The computation of TP and FP for the other five recommended landmarks is the

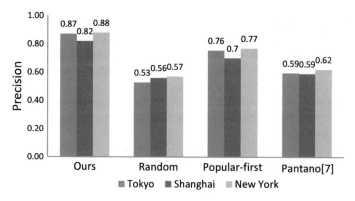

Fig. 6 Evaluation result on landmark recommendation precision

same. Then the *precision* for a user u can be obtained by:

$$Precision(u) = \frac{TP}{TP + FP} \tag{11}$$

Figure 6 shows the average precision of 12 users by our proposed algorithm, Random algorithm, Popular-first algorithm and the algorithm in [2]. Random algorithm has the lowest precision as it only randomly selects landmarks without considering the types of landmarks or user preferences. Popular-first algorithm recommends the top-6 ranked landmarks rated by a group of users in TripAdvisor. This algorithm improves precision as it considers users' general preferences. However, because this algorithm does not take the user's personalized preferences into consideration and hence our proposed algorithm's precision is around 10% higher than that of the Popular-first algorithm.

In [2], although it considers users' personalized preferences, it has poor performances with only around 60% precision. Because it is restricted to a small sample size of $N = 500$, as we discussed in Sect. 1 and it fails to resolve the problem of medium satisfaction (1, 2, 3 and 4 points). Pantano et al. assume that a user only has the very dislike (0 points) and the very like (5 points). But the truth is that the ratio of very dislike (0 points) cases in our landmark database rarely exists, which means 0 point cases should be considered less important compared with the medium satisfaction groups. Thus medium preferences should be considered rather than dividing users' preferences into only two extreme negative or positive groups.

7.2 Landmark Satisfaction Prediction Accuracy

We evaluate if our proposed accurately predicts the user's real satisfaction. The three lists, each of which contains 30 landmarks randomly chosen from each city database, are prepared and the users were required to write their real satisfaction on the landmarks in the list which they have visited before as many as possible. A user's real satisfaction on a landmark l is denoted as $^{real}S_{u,l}$. Through the questionnaires, 30 valid $^{real}S_{u,l}$ values were used as the ground-truths for each city.

We compare our proposed algorithm with the algorithms of the MAP (moving average predictor) that represents the users' average satisfaction [11] and the AVMAP (a variation moving average predictor) that represents a landmark's expectation satisfaction computed by the Dirichlet distribution [12]. Both the MAP and the AVMAP use a single data source.

The prediction precision is analyzed by (a) average *error rate*, (b) maximum error rate $(-/+)$ and (c) standard deviation (SD). *Error rate ER* is computed by:

$$ER = \frac{|^{pre}S_{u,l} - {}^{real}S_{u,l}|}{^{real}S_{u,l}} \times 100\% \tag{12}$$

where $^{pre}S_{u,l}$ is the predicted satisfaction of a user u on a landmark l by our algorithm, the MAP, or the AVMAP, and $^{real}S_{u,l}$ is a user u's the real satisfaction on a landmark l.

Table 3 lists the comparison results. In Table 3, it indicates that the proposed algorithm has the lowest error rate around 7% which is better than the MAP and the AVMAP, where only a single travel website comment data is used. This indicates that

Table 3 Evaluation result on satisfaction prediction accuracy

	Average error rate (%)	Maximum error rate $(-)$ (%)	Maximum error rate $(+)$ (%)	SD
Tokyo				
Ours	6.88	−12.80	11.90	0.08
MAP [11][a]	10.94	−26.00	18.42	0.12
MAP [11][b]	11.68	−20.00	18.42	0.13
AVMAP [12][a]	11.67	−30.00	20.00	0.13
AVMAP [12][b]	10.55	−18.40	14.75	0.11
Shanghai				
Ours	7.42	−14.75	11.90	0.10
MAP [11][a]	11.97	−30.00	20.75	0.13
MAP [11][a]	9.72	−20.00	15.75	0.11
AVMAP [12][a]	10.58	−30.00	20.97	0.13
AVMAP [12][b]	11.67	−20.00	18.42	0.14
New York				
Ours	7.47	−10.00	11.90	0.08
MAP [11][a]	10.63	−24.00	9.67	0.10
MAP [11][b]	11.43	−20.00	12.50	0.12
AVMAP [12][a]	9.46	−30.00	14.75	0.11
AVMAP [12][b]	10.05	−20.00	10.53	0.10

[a] Used comments from Ctrip only
[b] Used comments from TripAdvisor only

our proposed algorithm is effective in predicting users' satisfaction on landmarks in any cases and it is necessary to use multiple travel website data for better prediction accuracy.

Generally speaking, our proposed algorithm is very effective in predicting users' satisfaction on landmarks.

8 Conclusion and Future Works

We propose an algorithm that can effectively predict user's satisfaction on a landmark by the user's preferences on landmark type, language and travel websites. The results demonstrate that our proposed algorithm has a high degree of precision in terms of the landmark recommendation and landmark satisfaction prediction compared with previous studies.

In the future, a system for personalized travel route recommendation will be developed.

Acknowledgements This paper was supported in part by Grant-in-Aid for Scientific Research (No. 17K19986).

References

1. Filieri, R., Alguezaui, S., McLeay, F.: Why do travelers trust TripAdvisor? Antecedents of trust towards consumer-generated media and its influence on recommendation adoption and word of mouth. Tour. Manag. **51**, 174–185 (2015)
2. Pantano, E., Priporas, C., Stylos, N.: You will like it! using open data to predict tourists' response to a tourist attraction. Tour. Manag. **60**, 430–438 (2017)
3. Weaver, D., Kwek, A., Wang, Y.: Cultural connectedness and visitor segmentation in diaspora Chinese tourism. Tour. Manag. **63**, 302–314 (2017)
4. Dolnicar, S., Grun, B.: Assessing analytical robustness in cross-cultural comparisons. Int. J. Cult. Tour. Hosp. Res. **1**, 140–160 (2009)
5. Ctrip. http://www.ctrip.com
6. Jaran. https://www.jaran.jp
7. 4travel. http://4travel.jp
8. TripAdvisor. https://www.tripadvisor.jp
9. Lops, P., de Gemmis, M., Semeraro, M.: Basics of content-based recommender systems. In: Ricci, F., Rokach, L., Shapira, B. (eds.) Recommender Systems Handbook, pp. 75–78. Springer, New York (2011)
10. Darlington, R., Hayes, A.: Nonlinear relationships. In: Little, T. (ed.) Regression Analysis and Linear Models Concepts, Applications, and Implementation, pp. 358–362. Guilford Press, New York (2016)
11. de Jong, P.: Time series analysis. In Frees, E., Derrig, R., Meyers, G. (eds.) Predictive Modeling Applications in Actuarial Science: Volume 1, Predictive Modeling Techniques, Applications, and Implementation, pp. 434–436. Cambridge University Press, London (2014)
12. Khopkar, S., Nikolaev, A.: Predicting long-term product ratings based on few early ratings and user base analysis. Electron. Commer. Res. Appl. **21**, 38–49 (2017)

Joint Radio Resource Allocation in LTE-A Relay Networks with Carrier Aggregation

Jichiang Tsai

Abstract An LTE-A relay network constituted by an eNB and relay nodes is a cost-efficient way to extend the service coverage area of the eNB for providing higher bandwidth and better quality to users. However, one major issue arising with such a network is the potential interference among adjacent relay nodes, even between an RN and the eNB. Hence, how to address this issue to increase the overall system capacity becomes a big challenge. The downlink resource allocation problem in LTE-A relay networks has been widely studied in the literature to this end. Yet, these previous works have not considered that the scheduler can reassign Component Carriers to a User Equipment at any Transmission Time Interval. Furthermore, they have not considered the Modulation and Coding Scheme constraint specified in LTE-A standards. Particularly, these existing schemes can be only applied to systems without Carrier Aggregation configuration. In this paper, we introduce an efficient greedy scheduling algorithm to perform joint downlink radio resource allocation in an LTE-A outband relay network, additionally taking the above issues into consideration. Our proposed technique can also apply to the upcoming 5G system.

Keywords LTE-A networks · Resource allocation · Relay · Carrier aggregation
Proportional fairness

1 Introduction

The Long Term Evolution-Advanced (LTE-A) standard has proposed the Carrier Aggregation (CA) technique to accomplish wider bandwidth [8]. Through CA, up to five Component Carriers (CCs) can be aggregated to support high data rate transmission. CA allows multiple CCs to be utilized simultaneously as if it was one wide carrier employed for transmission. However, exploiting CA in the LTE-A network requires modification to the Radio Resource Management (RRM) entity, including

J. Tsai (✉)
Department of Electrical Engineering, National Chung Hsing University, Taichung 40227,
Taiwan, ROC
e-mail: jichiangt@nchu.edu.tw

© Springer Nature Switzerland AG 2019
R. Lee (ed.), *Computer and Information Science*, Studies in Computational
Intelligence 791, https://doi.org/10.1007/978-3-319-98693-7_9

CC selection, Resource Block (RB) allocation and Modulation and Coding Scheme (MCS) assignment.

The problem of downlink radio resource allocation for the LTE-A system with CA configuration has been explored in [15, 16]. These existing methods are mainly to design a load balancing mechanism to assign CCs to User Equipments (UEs) and then schedule RBs of CCs to the corresponding UEs at every Transmission Time Interval (TTI) to optimize the radio resource usage of the entire system. Yet, they have not considered that CCs assigned to a UE can be continually changed based on channel quality. This may confine the radio resource usage whenever the traffic loads and/or the channel conditions vary. Furthermore, one more constraint of the LTE system requires that only one MCS can be selected for each CC across all its RBs assigned to the same UE at a TTI in absence of Multiple Input Multiple Output (MIMO) spatial multiplexing [4]. But the aforementioned schemes all suppose that a UE can select two or more MCSs for different RBs of a CC assigned to it. Though the scheduling problem with the MCS constraint for LTE systems has been tackled in the literature [6, 9]. These corresponding schemes are dedicated to systems without CA configuration. Recently, an NP-hard problem for the downlink radio resource scheduling in LTE-A systems with CA configuration by further considering the foregoing issues has been formulated and solved in [13]. The proposed greedy scheduling algorithm has been shown to be able to provide at least half the performance of the optimal solution. Moreover, simulation results have demonstrated that such a scheme outperforms other corresponding schemes in the literature.

On the other hand, the LTE-A standard also supports the possibility of deploying relay nodes (RNs) within an E-UTRAN NodeB (eNB), termed donor eNB (DeNB) [1], as a cost-efficient way to improve the cell-edge throughput [12]. More explicitly, every RN accesses the DeNB through a wireless backhaul link (BL) and forwards data to and from some UE through a wireless access link (AL). Furthermore, only non-transparent relays are specified. That is, towards UEs, an RN appears like a regular eNB; while from an eNB perspective, it is viewed as a UE with additional features. Two main types of RNs are possible: In-band relays, if the backhaul and access links operate on the same carrier, and out-band relays, if the above two links operate on different carriers. With the deployment of RNs, a UE can obtain a larger channel gain and thus a better quality of communication as if it became closer to the DeNB. Yet, one of the major concerns arising with the LTE-A relay network is the potential mutual interference caused between two adjacent RNs, even between an RN and the DeNB. Thus how to mitigate such interference to increase the entire system capacity becomes a big challenge in the radio resource management of LTE-A relay networks [14].

The downlink resource allocation problem in LTE-A relay networks has been widely studied to maximize the overall system capacity [10, 14]. Still, these existing schemes have not considered that the scheduler can reassign CCs to any UE at each TTI and a sole MCS can be only adopted for all RBs of each CC assigned to a UE at any TTI. Moreover, they are merely applicable to systems without CA configuration. In this paper, we extend the results of the study in [13] to efficiently allocate downlink radio resources to all UEs in an LTE-A outband relay network with CA configuration and formulate the above task as a more complex optimization problem by further considering the issues caused by multiple relays. A greedy scheduling algorithm to this end is introduced as well. More importantly, the technique proposed in this paper can also apply to the forthcoming 5G system.

The rest of this paper is organized as follows: In Sect. 2, we describe the downlink transmission model of the LTE-A outband relay network and then formulate the task of its radio resource allocation as an optimization problem in Sect. 3. In Sect. 4, we introduce our greedy scheduling algorithm for the foregoing task. At last, we make a summary and provide some potential future work in Sect. 5.

2 Preliminaries

In this paper, we consider the downlink transmission in the LTE-A relay network. In such a network, there exist one DeNB, n RNs and m active UEs in the specified area, where $n, m \geq 1$, as illustrated by Fig. 1. Particularly, an LTE-A radio frame lasts 10 ms and is equally divided into 10 sub-frames. Each sub-frame of 1 ms duration is composed of two equal-sized time slots with 0.5 ms duration. The fundamental scheduling unit in LTE-A is an RB, which exactly occupies a time slot [1, 2]. Thus

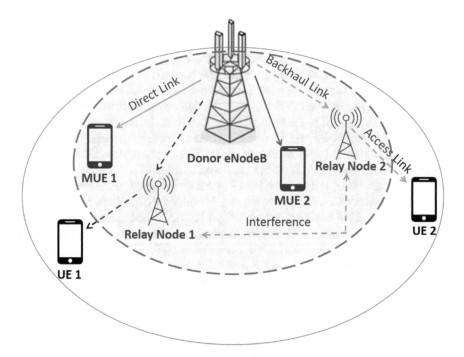

Fig. 1 The downlink scenario in an LTE-A relay network

the LTE-A downlink scheduling is the task of allocating RBs of CCs to UEs, which is performed at every TTI, i.e., a sub-frame duration. Any RB of a CC can be only assigned to a unique UE in absence of MIMO. Furthermore, with the CA technique, RBs of multiple CCs can be simultaneously allocated to a single UE at each TTI. In this paper, we assume that every CC has the same bandwidth of p RBs.

In the downlink transmission of the LTE-A outband relay network, an RN can receive data from the DeNB for UEs and meantime transmit data to UEs at the same TTI, by employing two distinct CCs. A UE can be exclusively served by the DeNB or some RN. The UEs directly served by the DeNB via the *direct link* are called *Macrocell User Equipments (MUEs)*. Moreover, a UE can move within the area such that the base station (the DeNB or an RN) serving it can be continually changed. The DeNB can always employ all CCs to transmit data while an RN can only use some of CCs to forward data. Here, we assume that the maximum number of CCs able to be used by an outband RN a is f_a, where $f_a \geq 2$. Similarly, each UE u is supposed to be able to employ at most $k_u \geq 1$ CC(s) to receive data. In this paper, we consider the backlogged traffic model, which means that the DeNB always has data for transmission to every UE at each TTI.

Since the channel conditions may vary with time, the frequency of a CC and the location of a UE, the channel condition of an RB of every CC for a UE is time-varying. In LTE, a UE can estimate individually the channel quality on each RB of all CCs by inspecting reference signals transmitted from the eNB [1]. Subsequently, the feedback report of the estimated channel quality is sent by each UE to the eNB in form of the Channel Quality Indicator (CQI), the value of which can then be mapped to the highest-rate Modulation and Coding Scheme (MCS) able to be adopted by the UE for receiving the corresponding RB from the eNB [3]. Likewise, in the LTE-A relay network, an RN needs to transmit its CQIs back to the DeNB for deciding the highest-rate MCSs that can be used by the former to receive RBs from the latter. Also, UEs have to send their CQIs to every RN that can serve them. In the text, our schedulers are supposed to know in advance the channel conditions on all RBs of CCs from all base stations to all their serving UEs and/or RNs according to the CQIs reported by the UEs and RNs. Moreover, the DeNB is connected to each its auxiliary RNs via a backhaul network. So all base stations can share Channel State Information (CSI) with each other, and then the DeNB and RNs can jointly schedule RBs of all CCs to UEs at every TTI. Finally, we assume that there are q available MCSs that can be employed by a UE or an RN, where MCS 1 has the lowest transmission rate and MCS q the highest transmission rate. Here, the achieved transmission rate on an RB with MCS ℓ from any RN or the DeNB is denoted as r_ℓ, the index of the highest-rate MCS able to be used by RN a on RB b of CC c from the DeNB is $Q_{a,c,b}$, and the index of the highest-rate MCS able to be used by UE u on RB b of CC c from RN a or the DeNB ($a = 0$) is $Q_{u,c,b,a}$. Obviously, the achieved transmission rate $v_{a,c,b,\ell}$ for RN a on RB b of CC c with MCS ℓ from the DeNB is r_ℓ if $\ell \leq Q_{a,c,b}$ and 0 otherwise. Similarly, the achieved transmission rate $v_{u,c,b,\ell,a}$ for UE u on RB b of CC c with MCS ℓ from station a is r_ℓ if $\ell \leq Q_{u,c,b,a}$ and 0 otherwise.

In a multi-transmitter environment, like the LTE-A relay network, we need to consider the issue of interference control. Particularly, each RN and the DeNB must transmit adequate power to provide coverage to its service area. Hence, if neighboring base stations share a channel at the same time, high interference will be induced. Obviously, such interference will dominate the performance of a base station and affect the precision of estimated channel qualities on RBs reported by an RN that depend on the radio resources actually employed by the neighboring stations of the RN later. So as to facilitate the interference control among base stations and improve the accuracy of feedback COIs from RNs, we make an orthogonal RB allocation constraint among neighboring base stations. In other words, a station cannot reuse the same RB if any of its neighboring station has used such an RB at a TTI. Note that if two base stations are geographically far enough such that the mutual interference between them can be ignored, they are allowed to be allocated with the same RB at every TTI. Therefore, we also assume that our schedulers have learned the set $\mathcal{N}_a$ of neighboring RNs of RN a that can cause interference on a. Note that $\mathcal{N}_0$ is the interference set of the DeNB. Because any RN can communicate with the DeNB, the two stations will interfere with each other mutually. Accordingly, the interference set $\mathcal{N}_0$ of the DeNB includes all RNs.

3 Resource Allocation Problem Formulation

We list in Table 1 some new symbols used in the text. For performing the resource allocation at every TTI, the scheduler need to find the values of the last two indicator variables $x_{u,c,b,\ell,a} \in \{0, 1\}$ and $y_{u,c,b,\ell,a} \in \{0, 1\}$, $\forall u \in \mathcal{U}$, $\forall c \in \mathcal{C}$, $\forall b \in \mathcal{B}$, $\forall \ell \in \mathcal{L}$, $\forall a \in \mathcal{R}$. More specifically, the goal of the proportional fair scheduler in question is to maximize the utility function $\sum_{\forall u \in \mathcal{U}} \log \mu_u(t)$, where $\mu_u(t)$ is the average transmission rate assigned by the eNB to UE u up to TTI t. Such a utility function is also known as *the proportional fair criterion*, and maximizing its value can

Table 1 A list of some new symbols used in the text

$\mathcal{A}$	Index set of every base station a, including all RNs and the DeNB ($a = 0$)
$\mathcal{R}$	Index set of every RN a
$\mathcal{U}$	Index set of every UE u
$\mathcal{C}$	Index set of every CC c
$\mathcal{B}$	Index set of every RB b per CC
$\mathcal{L}$	Index set of every MCS ℓ per RB
$\mathcal{S}_a$	Index set of UEs currently able to be served by station a
$\mathcal{U}_u$	Index set of stations by which UE u can be currently served
$x_{u,c,b,\ell,a}$	Indicator variable to denote if RB b of CC c with MCS ℓ from the DeNB is assigned to UE u via RN a (or directly from the DeNB if $a = 0$)
$y_{u,c,b,\ell,a}$	Indicator variable to denote if RB b of CC c with MCS ℓ from RN a is assigned to UE u

achieve high throughput of the whole system while maintaining proportional fairness of radio resource allocation among all UEs [5, 11]. In particular, it has been proved in the above two papers that the scheduler can maximize this utility function by maximizing the objective function $\sum_{\forall u \in \mathcal{U}} x_u(t) R_u(t)/\mu_u(t)$, where $x_u(t)$ is an indicator variable to denote if UE u is scheduled with radio resources at TTI t and $R_u(t)$ the current total transmission rate assigned by the eNB to UE u at TTI t. In the above objective function, $1/\mu_u(t)$ can be viewed as the priority weight w_u of UE u on resource allocation at TTI t in the sense that a UE with a lower priority weight, i.e., with a lower average transmission rate assigned from the eNB, in the past will enjoy higher priority being assigned with more resources in the current TTI to maintain proportional fairness. Now that an RN does not connect to a fixed network and its purpose is wirelessly relaying data to UEs, it cannot actually supply any additional bandwidth, and thus extra utility, to UEs. So we formulate our scheduling problem of allocating radio resources to UEs in the LTE-A relay network as maximizing the weighted sum of transmission rates of UEs assigned from the DeNB at TTI t, under the constraints conforming to the specification of the outband RN. To simplify notation, the TTI index t is omitted. Then, our scheduling problem is formulated as solving the optimization problem expressed below:

$$
\max \sum_{\forall u \in \mathcal{U}} w_u \times \left(\sum_{\forall c \in \mathcal{C}} \sum_{\forall b \in \mathcal{B}} \sum_{\forall \ell \in \mathcal{L}} x_{u,c,b,\ell,0} \times v_{u,c,b,\ell,0} \right.
$$

$$
\left. + \sum_{\forall a \in \mathcal{R}} \sum_{\forall c \in \mathcal{C}} \sum_{\forall b \in \mathcal{B}} \sum_{\forall \ell \in \mathcal{L}} x_{u,c,b,\ell,a} \times v_{a,c,b,\ell} \right) \tag{1}
$$

subjected to the following constraints:

$$
\sum_{\forall u \in \mathcal{U}} \sum_{\forall \ell \in \mathcal{L}} \sum_{\forall a \in \mathcal{A}} x_{u,c,b,\ell,a} \leq 1, \forall c \in \mathcal{C}, \forall b \in \mathcal{B} \tag{2}
$$

$$
\sum_{\forall u \in \mathcal{U}} \sum_{\forall \ell \in \mathcal{L}} y_{u,c,b,\ell,a} \leq 1, \forall c \in \mathcal{C}, \forall b \in \mathcal{B}, \forall a \in \mathcal{R} \tag{3}
$$

$$
\sum_{\forall \ell \in \mathcal{L}} \max_{\forall b \in \mathcal{B}} x_{u,c,b,\ell,0} \leq 1, \forall u \in \mathcal{U}, \forall c \in \mathcal{C} \tag{4}
$$

$$
\sum_{\forall \ell \in \mathcal{L}} \max_{\forall u \in \mathcal{U}} \max_{\forall b \in \mathcal{B}} x_{u,c,b,\ell,a} \leq 1, \forall c \in \mathcal{C}, \forall a \in \mathcal{R} \tag{5}
$$

$$
\sum_{\forall \ell \in \mathcal{L}} \max_{\forall b \in \mathcal{B}} y_{u,c,b,\ell,a} \leq 1, \forall u \in \mathcal{U}, \forall c \in \mathcal{C}, \forall a \in \mathcal{R} \tag{6}
$$

$$
\sum_{\forall c \in \mathcal{C}} \sum_{\forall b \in \mathcal{B}} \sum_{\forall \ell \in \mathcal{L}} x_{u,c,b,\ell,a} \times \sum_{\forall c \in \mathcal{C}} \sum_{\forall b \in \mathcal{B}} \sum_{\forall \ell \in \mathcal{L}} \sum_{\forall a' \in \mathcal{A} -\{a\}} x_{u,c,b,\ell,a'} = 0,
$$

$$
\forall u \in \mathcal{U}, \forall a \in \mathcal{A} \tag{7}
$$

$$\sum_{\forall c \in \mathscr{C}} \sum_{\forall b \in \mathscr{B}} \sum_{\forall \ell \in \mathscr{L}} y_{u,c,b,\ell,a} \times \sum_{\forall c \in \mathscr{C}} \sum_{\forall b \in \mathscr{B}} \sum_{\forall \ell \in \mathscr{L}} \sum_{\forall a' \in \mathscr{R} - \{a\}} y_{u,c,b,\ell,a'} = 0,$$
$$\forall u \in \mathscr{U}, \forall a \in \mathscr{R} \quad (8)$$

$$\max_{\forall c \in \mathscr{C}} \max_{\forall b \in \mathscr{B}} \max_{\forall \ell \in \mathscr{L}} x_{u,c,b,\ell,a} = \max_{\forall c \in \mathscr{C}} \max_{\forall b \in \mathscr{B}} \max_{\forall \ell \in \mathscr{L}} y_{u,c,b,\ell,a}, \forall u \in \mathscr{U}, \forall a \in \mathscr{R} \quad (9)$$

$$\sum_{\forall u \in \mathscr{U}} \sum_{\forall \ell \in \mathscr{L}} x_{u,c,b,\ell,0} \times \sum_{\forall u \in \mathscr{U}} \sum_{\forall \ell \in \mathscr{L}} y_{u,c,b,\ell,a} = 0, \forall c \in \mathscr{C}, \forall b \in \mathscr{B}, \forall a \in \mathscr{R} \quad (10)$$

$$\sum_{\forall u \in \mathscr{U}} \sum_{\forall \ell \in \mathscr{L}} y_{u,c,b,\ell,a} \times \sum_{\forall u \in \mathscr{U}} \sum_{\forall \ell \in \mathscr{L}} \sum_{\forall a' \in \mathscr{N}_a} y_{u,c,b,\ell,a'} = 0,$$
$$\forall c \in \mathscr{C}, \forall b \in \mathscr{B}, \forall a \in \mathscr{R} \quad (11)$$

$$\sum_{\forall u \in \mathscr{U}} \sum_{\forall \ell \in \mathscr{L}} x_{u,c,b,\ell,a} \times \sum_{\forall u \in \mathscr{U}} \sum_{\forall \ell \in \mathscr{L}} \sum_{\forall a' \in \mathscr{N}_a} y_{u,c,b,\ell,a'} = 0,$$
$$\forall c \in \mathscr{C}, \forall b \in \mathscr{B}, \forall a \in \mathscr{R} \quad (12)$$

$$\sum_{\forall u \in \mathscr{U}} \sum_{\forall b \in \mathscr{B}} \sum_{\forall \ell \in \mathscr{L}} x_{u,c,b,\ell,a} \times \sum_{\forall u \in \mathscr{U}} \sum_{\forall b \in \mathscr{B}} \sum_{\forall \ell \in \mathscr{L}} y_{u,c,b,\ell,a} = 0,$$
$$\forall c \in \mathscr{C}, \forall a \in \mathscr{R} \quad (13)$$

$$\sum_{\forall c \in \mathscr{C}} \max_{\forall u \in \mathscr{U}} \max_{\forall b \in \mathscr{B}} \max_{\forall \ell \in \mathscr{L}} (x_{u,c,b,\ell,a} + y_{u,c,b,\ell,a}) \leq f_a, \forall a \in \mathscr{R} \quad (14)$$

$$\sum_{\forall c \in \mathscr{C}} \max_{\forall b \in \mathscr{B}} \max_{\forall \ell \in \mathscr{L}} \max_{\forall a \in \mathscr{R}} (x_{u,c,b,\ell,0} + y_{u,c,b,\ell,a}) \leq k_u, \forall u \in \mathscr{U} \quad (15)$$

$$\sum_{\forall u \in \mathscr{U}} \sum_{\forall \ell > Q_{a,c,b}} x_{u,c,b,\ell,a} = 0, \forall c \in \mathscr{C}, \forall b \in \mathscr{B}, \forall a \in \mathscr{R} \quad (16)$$

$$\sum_{\forall \ell > Q_{u,c,b,0}} x_{u,c,b,\ell,0} = 0, \forall u \in \mathscr{U}, \forall c \in \mathscr{C}, \forall b \in \mathscr{B} \quad (17)$$

$$\sum_{\forall \ell > Q_{u,c,b,a}} y_{u,c,b,\ell,a} = 0, \forall u \in \mathscr{U}, \forall c \in \mathscr{C}, \forall b \in \mathscr{B}, \forall a \in \mathscr{R} \quad (18)$$

$$\sum_{\forall u \in \mathscr{U} - \mathscr{S}_a} \sum_{\forall c \in \mathscr{C}} \sum_{\forall b \in \mathscr{B}} \sum_{\forall \ell \in \mathscr{L}} x_{u,c,b,\ell,a} = 0, \forall a \in \mathscr{A} \quad (19)$$

$$\sum_{\forall u \in \mathscr{U} - \mathscr{S}_a} \sum_{\forall c \in \mathscr{C}} \sum_{\forall b \in \mathscr{B}} \sum_{\forall \ell \in \mathscr{L}} y_{u,c,b,\ell,a} = 0, \forall a \in \mathscr{R} \quad (20)$$

$$\sum_{\forall c \in \mathscr{C}} \sum_{\forall b \in \mathscr{B}} \sum_{\forall \ell \in \mathscr{L}} x_{u,c,b,\ell,a} \times v_{a,c,b,\ell} \leq \sum_{\forall c \in \mathscr{C}} \sum_{\forall b \in \mathscr{B}} \sum_{\forall \ell \in \mathscr{L}} y_{u,c,b,\ell,a} \times v_{u,c,b,\ell,a},$$
$$\forall u \in \mathscr{U}, \forall a \in \mathscr{R} \quad (21)$$

First, the former term in the aforementioned objective function denotes the total UE utility contributed by all MUEs; while the second term means that contributed by all other UEs served by the DeNB via RNs. Next, the constraints (2) and (3) are to ensure that each RB of a CC from some RN or the DeNB can be only assigned to a UE or an RN for a sole UE with a certain MCS. As for the three succeeding constraints in inequalities (4), (5) and (6), they dictate that all RBs of a CC from some RN or the DeNB assigned to a UE or an RN for all its serving UEs can only adopt one unique MCS. Moreover, because each UE can be exclusively served by one base station without MIMO, the constrain in equality (7) guarantees that any UE can be served only by one RN or the DeNB, whereas that in (8) ensures that a UE that is not served directly by the DeNB can be merely served by one RN. Particularly, the subsequent constraint in equality (9) dictates that a UE must be served by the RN that receives data from the DeNB for the UE.

On the other hand, there are another four constrains ensuring that any RB of a CC cannot be recycled if there is any interference between two stations. More explicitly, constraint (10) guarantees that the DeNB and any RN cannot transmit the same RB of a CC, constraint (11) dictates that an RN cannot transmit the same RB as its neighboring RNs transmit, and constraint (12) means that an RN cannot transmit the same RB as its neighboring RNs receive. Furthermore, since we only consider the outband RN in this paper, the constraint in the following equality (13) restricts that every RN does not use the same CC for both transmission and reception at a TTI. As for the following two constraints, they simply mean that the total number of CCs employed by an RN or a UE, including an MUE, cannot exceed its limit. In addition, although the combinations of the two indicator variables $x_{u,c,b,\ell,a}$ and $y_{u,c,b,\ell,a}$ are numerous, the succeeding five constraints exhibit that values of many combinations are zero because for an RB of any CC, an RN or a UE does not select an MCS unable to be used by it (the three equalities from (16)) and the DeNB or an RN cannot serve any UE currently unable to be served by it (equalities (19) and (20)). At last, the final constraint in inequality (21) is to ensure that the transmission rate assigned by an RN to a UE is not less than that assigned by the DeNB to the UE through the RN. The purpose of such a constraint is twofold. First, as mentioned previously, we assume the backlogged model for our system that the DeNB always has data to transmit to a UE at each TTI. Hence, for the stability of the DeNB queues and RN queues, whenever an RN receives an amount of data form the DeNB for a UE at a TTI, it needs to forward the whole data to the UE at the same TTI. Second, now that the DeNB has allocated so much bandwidth to a UE via an RN, the RN had better assign at least the same amount of bandwidth to the UE for entirely consuming the bandwidth from the DeNB to preserve the efficiency of the scheduler.

Recently, a proportional-fair radio resource scheduling problem for the LTE-A system with the CA configuration under which the system enjoys multiple CCs and multiple MCSs for data transmission has been proposed in [13]. This new optimization problem is solved at each TTI subjected to the three constraints below: (i) Every RB is assigned to at most one UE; (ii) A UE can merely employ one MCS of multiple MCSs for all RBs of one of its assigned CCs; (iii) A UE can be simultaneously assigned with RBs of multiple CCs without exceeding its limit. It has come to a conclusion in [13] that the foregoing optimization problem is NP-hard.

In this paper, we further extend the work in [13] to proportional-fairly allocate radio resources to all UEs in the LTE-A outband relay network and formulate such a task as a much more complex optimization problem in the preceding paragraphs. It is not difficult to realize that when there is not any RN in the system, i.e., all UEs are MUEs, our problem can be reduced to the one introduced in [13]. Therefore, our optimization problem is still NP-hard.

4 The Greedy Resource Allocation Algorithm

The authors in [13] have proposed one efficient greedy scheduling algorithm to find a sub-optimal solution of their optimization problem. Based on a concept enhanced from the foregoing work, we also introduce a greedy scheduling algorithm for sub-optimally solving our optimization problem in this section. More specifically, the idea behind the greedy scheduling algorithm in [13] is to make the UE utility contributed by every RB from the eNB, i.e., the incremental weighted transmission rate of some UE due to being assigned with the RB, as large as possible. In doing so, the scheduler can properly allocate any RB from the eNB to a UE to maximize the overall utility of the system as it can, which is exactly the goal of the scheduler. As a result, the UE utility contributed by each RB of a CC from the eNB with any combination of UE allocation and MCS selection is first calculated. Moreover, an assignment of a CC from the eNB with a certain MCS to a UE is formulated as a gain, which is the maximal total utility increment to the system that can be achieved by reassigning some RBs of the CC to the UE. In particular, gains of all remaining potential assignments of UEs, CCs, and MCSs are computed in each iteration, and then the assignment with the highest nonzero gain is adopted iteratively until all UEs are allocated with CCs with suitable MCSs or no any better assignment with a higher gain can be found. Such a scheduling task is performed at every TTI.

Our scheduling algorithm is devised in light of a concept of greediness improved from the one mentioned previously. More explicitly, the algorithm proposed in [13] only considers assignments that allocate just one CC from the eNB to a UE in each iteration, whereas our algorithms consider those with up to two CCs from the DeNB directly or indirectly assigned to at most two UEs in an iteration. The reason behind the above action is that an outband RN needs two distinct carriers for forwarding data to a UE: one for the backhaul link and the other for the access link. Hence, so as to maximize the total UE utility of the system as we can, we need to take all possible assignments of up to two CCs from the DeNB directly to at most two UEs and those of one CC from the DeNB indirectly to a UE through some RN along with the other CC from the RN to the UE into consideration. Doing so seems to favor MUEs because every RB from an RN does not really contribute any UE utility to the overall system. However, due to the proportional fairness imposed on the objective of our scheduler, as TTIs elapse, all UEs will be eventually allocated with almost the same amount of radio resources while achieving high system throughput.

4.1 Key Variables

First, we list all key variables used in our scheduling algorithm as well as their initial values and purposes in Fig. 2. Particularly, for the set $\mathcal{M}$ of all possible assignments, if $u_2 = 0$ (or $c_2 = 0$), there is no second UE (or CC) considered for this assignment. Moreover, for the case of being served by an RN, i.e., $a \neq 0$, we do not need the second UE but require two distinct CCs for data transmission. As for the case of MUEs, we need to consider the assignments of merely one CC to one sole UE, and if $u_1 = u_2$, it makes sense to allocate two different CCs to the same MUE. Subsequently, the utility value contributed by an RB of a CC, i.e., $h_D(u, c, b, \ell, a)$, is $w_u \times v_{u,c,b,\ell,0}$ if $a = 0$ and $w_u \times v_{a,c,b,\ell}$ otherwise. Correspondingly, the bandwidth contributed by RBs from the DeNB assigned to a UE via some RN must be accomplished by RBs from the RN for the UE. Hence, we also have to calculate the utility value that can be realized by an RB of a CC from an RN, i.e., $h_R(u, c, b, \ell, a)$. Its value is $w_u \times v_{u,c,b,\ell,a}$.

Next, for the gain $g(u_1, u_2, c_1, c_2, \ell_1, \ell_2, a)$ of an assignment, if $c_2 = 0$, no second MCS is required such that ℓ_2 is set to zero. Also, if $a \neq 0$, $\ell_2 = 0$ since the scheduler will find the most suitable MCS for the carrier from an RN to its serving UE. Finally, as mentioned previously, the base station serving a certain UE may alter due to the mobility of the UE. However, to avoid unacceptable handover overhead [7], we only

(01) $\mathcal{M} \leftarrow \{(u_1, u_2, c_1, c_2, a) \mid u_1 \in \mathcal{S}_a, u_2 \in \mathcal{S}_0 \cup \{0\}, c_1 \in \mathcal{C}, c_2 \in \mathcal{C} \cup \{0\}, a \in \mathcal{A}, a \neq 0 \rightarrow u_2 = 0 \wedge c_2 \neq 0 \wedge c_1 \neq c_2, a = 0 \wedge u_2 = 0 \rightarrow c_2 = 0, u_1 = u_2 \rightarrow c_1 \neq c_2\}$: set of all potential assignments of two CCs c_1, c_2 from the DeNB to two UEs u_1, u_2 and those of CC c_1 from the DeNB to UE u_1 via RN a along with CC c_2 from a to u_1;

(02) $\mathcal{H}(c, b, a) \leftarrow 0, \forall c \in \mathcal{C}, \forall b \in \mathcal{B}, \forall a \in \mathcal{A}$: current utility value contributed by RB b of CC c from the DeNB perceived by an RN or the DeNB;

(03) $h_D(u, c, b, \ell, a), \forall u \in \mathcal{S}_a, \forall c \in \mathcal{C}, \forall b \in \mathcal{B}, \forall \ell \in \mathcal{L}, \forall a \in \mathcal{A}$: utility value contributed by RB b of CC c with MCS ℓ from the DeNB assigned to UE u via RN a (or directly from the DeNB if $a = 0$);

(04) $h_R(u, c, b, \ell, a), \forall u \in \mathcal{S}_a, \forall c \in \mathcal{C}, \forall b \in \mathcal{B}, \forall \ell \in \mathcal{L}, \forall a \in \mathcal{R}$: utility value made by RB b of CC c with MCS ℓ from RN a assigned to UE u;

(05) $g(u_1, u_2, c_1, c_2, \ell_1, \ell_2, a), \forall (u_1, u_2, c_1, c_2, a) \in \mathcal{M}, \forall \ell_1 \in \mathcal{L}, \forall \ell_2 \in \mathcal{L} \cup \{0\}$: gain of the assignment (u_1, u_2, c_1, c_2, a) with CCs c_1, c_2 adopting MCSs ℓ_1, ℓ_2, respectively;

(06) $\ell_{a,c}^D \leftarrow 0, \forall c \in \mathcal{C}, \forall a \in \mathcal{R}$: index of the current MCS used by CC c from the DeNB assigned to RN a;

(07) $\ell_{u,c}^R \leftarrow 0, \forall u \in \mathcal{U}, \forall c \in \mathcal{C}$: index of the current MCS used by CC c from some RN assigned to UE u;

(08) $\mathcal{V}_D(u) \leftarrow \emptyset, \forall u \in \mathcal{U}$: set of pairs (c, b) with RB b of CC c from the DeNB assigned to UE u via some RN or directly from the DeNB;

(09) $\mathcal{V}_R(u) \leftarrow \emptyset, \forall u \in \mathcal{U}$: set of pairs (c, b) with RB b of CC c from some RN assigned to UE u;

(10) $\mathcal{K}_D(u) \leftarrow 0, \forall u \in \mathcal{U}$: current total utility value contributed by UE u on RBs from the DeNB via an RN;

(11) $\mathcal{K}_R(u) \leftarrow 0, \forall u \in \mathcal{U}$: current total utility value accomplished by UE u on RBs from an RN;

(12) $\mathcal{S}_a^t \leftarrow \emptyset, \forall a \in \mathcal{R}$: index set of UEs currently served by RN a or the DeNB if $a = 0$;

Fig. 2 The key variables employed in the scheduling algorithm

allow a UE to change its station every moderate period, say 100 TTIs. Therefore, both $\mathscr{S}_a$ and $\mathscr{Z}_u$ presented in Table 1 can be updated merely at the very beginning of the above pre-defined period. Furthermore, after the first TTI of the period, $\mathscr{S}_a$ is set to $\mathscr{S}_a^t$, the set of UEs currently served by station a, obtained by the end of such a TTI and left the same during the remaining TTIs of the same period. Likewise, $\mathscr{Z}_u$ is set to the station assigned to UE u at the first TTI of the pre-defined period and kept unchanged by the end of the period.

4.2 Calculating Gains of Possible Assignments

The core task of our scheduler is to calculate the gain $g(u_1, u_2, c_1, c_2, \ell_1, \ell_2, a)$ of each potential assignment (u_1, u_2, c_1, c_2, a) adopting MCSs ℓ_1 and ℓ_2 for its two CCs, presented as Subroutine_1 in Fig. 3. First, for an assignment $(u_1, u_2, c_1, c_2, 0)$ of up to two CCs c_1, c_2 to at most two MUEs u_1, u_2, the part of its gain contributed by the first MUE u_1, namely, the maximal total utility increment to the system that can be achieved by reassigning some RBs of c_1 to u_1, is obtained in lines 1–2 with the same way as that in [13]. Likewise, if the second MUE u_2 is involved, i.e., $u_2 \neq 0$, the part of the gain contributed by u_2 is also calculated in a similar way and then added to $g(u_1, u_2, c_1, c_2, \ell_1, \ell_2, a)$ (lines 3–4). Note that when $u_1 = u_2$, i.e., the considered

---

(01) **if** $a = 0$ **then**
(02) $g(u_1,u_2,c_1,c_2,\ell_1,\ell_2,a) \leftarrow \sum_{\forall b \in \mathscr{B}} \max(0, h_D(u_1,c_1,b,\ell_1,0) - \mathscr{H}(c_1,b,0));$
(03)   **if** $u_2 \neq 0 \wedge$ (assigning c_2 to UE u_2 will not exceed k_{u_2} CCs) **then**
(04) $g(u_1,u_2,c_1,c_2,\ell_1,\ell_2,a) \leftarrow g(u_1,u_2,c_1,c_2,\ell_1,\ell_2,a)$
         $+ \sum_{\forall b \in \mathscr{B}} \max(0, h_D(u_2,c_2,b,\ell_2,0) - \mathscr{H}(c_2,b,0));$
(05)   **else if** $u_2 = u_1$ **then** $g(u_1,u_2,c_1,c_2,\ell_1,\ell_2,a) \leftarrow 0;$
(06) **else**
(07) $g(u_1,u_2,c_1,c_2,\ell_1,\ell_2,a) \leftarrow 0;$
(08)   **if** (RN a using both c_1 and c_2 will not exceed f_a CCs) $\wedge$ $(\ell_{a,c_1}^D = 0 \vee \ell_1 = \ell_{a,c_1}^D)$ **then**
(09) $\mathscr{D}^t \leftarrow$ RB(s) b of CC c_1 with $h_D(u_1,c_1,b,\ell_1,a) > \mathscr{H}(c_1,b,a);$
(10)   Sort RB(s) b of $\mathscr{D}^t$ in the descending order of $(h_D(u_1,c_1,b,\ell_1,a) - \mathscr{H}(c_1,b,a));$
(11)   **for each** $n \in [1, |\mathscr{D}^t|]$ **do**
(12) $\mathscr{D}^w \leftarrow$ first n RB(s) in $\mathscr{D}^t; g^t \leftarrow \sum_{\forall b \in \mathscr{D}^w}(h_D(u_1,c_1,b,\ell_1,a) - \mathscr{H}(c_1,b,a));$
(13) $\mathscr{K} \leftarrow \max(0, \sum_{\forall b \in \mathscr{D}^w} h_D(u_1,c_1,b,\ell_1,a) - (\mathscr{K}_R(u_1) - \mathscr{K}_D(u_1)));$
(14)   **for** $\forall \ell \in \mathscr{L} \wedge \sum_{\forall b \in \mathscr{B} \wedge \mathscr{H}(c_2,b,a) \neq \infty} h_R(u_1,c_2,b,\ell,a) \geq \mathscr{K}$ **do**
(15)   **if** $\ell_{u_1,c_2}^R = 0 \vee \ell = \ell_{u_1,c_2}^R$ **then**
(16) $\mathscr{E} \leftarrow$ all RB(s) b of CC c_2 with $h_R(u_1,c_2,b,\ell,a) > 0 \wedge \mathscr{H}(c_2,b,a) \neq \infty;$
(17)   Sort RB(s) b of $\mathscr{E}$ in the ascending order of $\mathscr{H}(c_2,b,a);$
(18)   Keep the first $\lceil \frac{\mathscr{K}}{w_{u_1} \times r_\ell} \rceil$ RB(s) in $\mathscr{E}$ and discard others;
(19) $g^t \leftarrow \max(0, g^t - \sum_{\forall b \in \mathscr{E}} \mathscr{H}(c_2,b,a));$
(20)   **if** $g^t > g(u_1,u_2,c_1,c_2,\ell_1,\ell_2,a)$ **then**
(21) $g(u_1,u_2,c_1,c_2,\ell_1,\ell_2,a) \leftarrow g^t; \ell^* \leftarrow \ell; \mathscr{D}^* \leftarrow \mathscr{D}^w; \mathscr{E}^* \leftarrow \mathscr{E};$

Fig. 3 Subroutine_1: Calculating the value of $g(u_1, u_2, c_1, c_2, \ell_1, \ell_2, a)$

two MUEs are the same, it is necessary to check if assigning two more CCs c_1, c_2 to such an MUE will not exceed the limit of the MUE on number of usable CCs (line 3). If not, this assignment is not allowed so that its gain is set to zero (line 5).

In contrast, for the assignment $(u_1, 0, c_1, c_2, a)$ of CC c_1 with MCS ℓ_1 from the DeNB to UE u_1 indirectly via RN a and CC c_2 with MCS ℓ_2 from a to u_1, the procedure for calculating the gain of this assignment is more sophisticated than the one proposed in [13]. We initially set the gain to zero (line 7). Then we continue to verify the validness of this assignment by determining two conditions in line 8: (i) RN a employing both CCs c_1 and c_2 will not exceed f_a CCs; (ii) CC c_1 from the DeNB has not been yet allocated to a for any UE, i.e., the index of the current MCS used by c_1 for a is zero ($\ell_{a,c_1}^D = 0$), or c_1 is assigned to a again with the same MCS as the current one ($\ell_1 = \ell_{a,c_1}^D$). The latter condition is required since our algorithm does not change the MCS of RBs that have been allocated to an RN. If either of the foregoing two conditions does not hold, such an assignment is not allowed and thus its gain remains zero.

If the assignment in question is valid, we find out any RB b of CC c_1 with MCS ℓ_1 that has a larger utility value for u_1 than that currently perceived by RN a (line 9), as well as sorting these RBs in the descending order of the utility increment contributed by an RB (line 10). Subsequently, unlike the assignment to an MUE, we cannot simply reassign all the above RBs to RN a for UE u_1 because with more bandwidth assigned to a UE via an RN, the RN may need to allocate more RBs to the UE for accomplishing such bandwidth and in turn may cause the DeNB to use less RBs due to the interference between base stations. The above situation may lead to the reduction of the total UE utility contributed by the system instead. As a result, we gradually add these eligible RBs of the same bandwidth one by one in the descending order of the corresponding utility increment to greedily search the best gain of this assignment (line 11). Then, we calculate ahead the total utility increment contributed by all selected RBs in $\mathscr{D}^w$ so as to decrease some computational complexity of the inner loop below (line 12). In addition, we compute the amount $\mathscr{K}$ of utility that needs to be additionally realized by a with the assignment of some RBs of CC c_2 to u_1 in line 13.

Next, in the inner loop (lines 14–21), we proceed with the assignment of RBs of CC c_2 from RN a to UE u_1. At this stage, we attempt to search a suitable MCS for c_2 to lower the total UE utility contributed by the whole system the least. Hence, we find out any MCS ℓ able to make the overall utility of all RBs of c_2 that have not been employed by a and its neighboring RNs complement the additional amount of utility required to be realized by a (line 14). Note that if no such an MCS exists, the gain of this assignment remains zero. Likewise, because the algorithm does not alter the MCS of RBs that have been allocated to a UE from an RN, it is also necessary to check that CC c_2 from RN a has not been yet allocated to UE u_1 or c_2 is assigned to u_1 again with the same MCS as the current one (line 15). If so, we find out all RBs that can accomplish some utility for u_1 and have not yet used by a and its neighboring RNs (line 16). Then, we sort these eligible RBs in $\mathscr{E}$ in the ascending order of the associated utility currently perceived by RN a (line 17). The purpose of the above action intends to reduce the current total UE utility of the system as less as possible due to reassigning some RBs of CC c_2 to u_1. Then, we re-allocate the first $\lceil \frac{\mathscr{K}}{w_{u_1} \times r_\ell} \rceil$ RBs of $\mathscr{E}$ to UE u_1 to accomplish the additional utility presented by $\mathscr{K}$ since the utility realized by any RB with MCS ℓ for u_1 is $w_{u_1} \times r_\ell$ (line 18). According to the foregoing RB re-allocation, we also calculate the gain of the assignment $(u_1, 0, c_1, c_2, a)$ with CC c_2 adopting MCS ℓ, namely, the overall utility increment contributed by the selected RBs of CC c_1 from the DeNB minus the total utility perceived by RN a of the selected RBs of CC c_2 from a in line 19. If the resultant gain is larger than the present maximum one obtained in an earlier iteration with another MCS (line 20), we update the corresponding information in line 21 to keep track of that for the largest gain found so far.

4.3 Releasing RBs from RNs

One more task of our scheduling algorithm is to release some RBs of RNs that have been allocated to realize the bandwidth from the DeNB. More explicitly, during the scheduling course at any TTI, the scheduler may direct the DeNB to reassign an RB b originally assigned to an RN a for a UE u to an MUE or any RN else. For the above situation, chances are that a can free some of its RBs originally allocated to u for accomplishing the bandwidth of b to allow any base station nearby, including itself and the DeNB, to employ these RBs for some other UEs. Such a task corresponding to CC c from the DeNB is performed by Subroutine_2 presented in Fig. 4. Note that in our scheduling algorithms, any RB used by an RN can be reused by any neighboring base station of the RN only after such an RB has released by the RN. First, for each RN a, if any RB of CC c originally assigned to a UE u being served by a is reassigned to another MUE or RN, we begin to release some RBs currently used by a for realizing the bandwidth of CC c allocated to u (lines 1–2). We recalculate

---

(01) **for each** $a \in \mathscr{R}$ **do**

(02)   **for each** $u \in \mathscr{S}_a^t$ with some pair (c,b) removed from $\mathscr{V}_D(u)$ **do**

(03) $\mathscr{K}_D(u) \leftarrow \mathscr{K}_D(u)-$ total utility value of UE u on removed (c,b);

(04)   Keep the minimum number of (c',b) in $\mathscr{V}_R(u)$ from the beginning
             with a total utility value $\geq \mathscr{K}_D(u)$ and discard others;

(05) $\mathscr{K}_R(u) \leftarrow \mathscr{K}_R(u)-$ total utility value of UE u on pairs (c',b) discarded from $\mathscr{V}_R(u)$;

(06)   **for each** $c' \in \mathscr{C}$ with some pair (c',b) discarded from $\mathscr{V}_R(u)$ **do**

(07)   **if** $\forall b \in \mathscr{B}, \mathscr{V}_R(u)$ does not contain (c',b) **then** $\ell_{u,c'}^R \leftarrow 0$;

(08)   **for each** pair (c',b) discarded from $\mathscr{V}_R(u)$ **and each** $a' \in \mathscr{N}_a \cup \{0,a\}$ **do**

(09)   **if** all RNs in $\mathscr{N}_{a'}$ do not use RB b **then** $\mathscr{H}(c',b,a') \leftarrow 0$;

(10)   **if** $\exists u \in \mathscr{S}_a^t, \exists b \in \mathscr{B}$, pair (c,b) has been removed from $\mathscr{V}_D(u)$ **then**

(11)   **if** $\forall u \in \mathscr{S}_a^t, \forall b \in \mathscr{B}, \mathscr{V}_D(u)$ does not contain (c,b) **then** $\ell_{a,c}^D \leftarrow 0$;

Fig. 4 Subroutine_2: releasing redundant RBs of RNs that originally accomplish the bandwidth of CC c from the DeNB

the present total utility value contributed by u on RBs from the DeNB via a in line 3. Then, we keep the minimum number of CC-RB pairs in $\mathcal{V}_R(u)$ that can accomplish the resultant utility for UE u and discard others (line 4). In line 5, we recalculate the current total utility value realized by u on RBs from RN a.

More importantly, in the following inner loop (lines 6–9), we proceed with freeing the above redundant RBs. Particularly, for each CC c' with some CC-RB pairs discarded from $\mathcal{V}_R(u)$ (line 6), we first check if no RB of c' is allocated to UE u. If so, we set $\ell_{u,c'}^R$ to zero (line 7). Furthermore, for any RB b of CC c' discarded from $\mathcal{V}_R(u)$, we decide if each base station a' nearby, including RN a and the DeNB, can employ b (lines 8–9). In other words, we check if all neighboring RNs of a' do not use RB b to allow such a base station to use b. At last, if there exists some RB of CC c originally allocated to RN a for some UE is re-allocated to another RN or MUE, we check if a has been unassigned with any RB of c or not for deciding if the value of $\ell_{a,c}^D$ is set to zero (lines 10–11).

4.4 The Scheduling Algorithm

Our greedy scheduling algorithm is presented in Fig. 5. In the very beginning, we calculate the values of $h_D(u, c, b, \ell, a)$ and $h_R(u, c, b, \ell, a)$ for every possible combination of UE u, CC c, RB b, MCS ℓ and base station a, respectively. In the following main loop (lines 3–39), from all assignments remaining in $\mathcal{M}$ with up to two arbitrary MCSs, we search the one enjoying the highest gain and then do the corresponding reassignment iteratively until all potential assignments have been removed from $\mathcal{M}$ or no assignment with a non-zero gain is available. We first calculate the gain of every possible assignment by Subroutine_1 to find the assignment $(u_1^*, u_2^*, c_1^*, c_2^*, \ell_1^*, \ell_2^*, a^*)$ with the largest gain (lines 4–5). If the gain of the above assignment is zero, the scheduling for the current TTI is terminated (lines 6 and 39). In contrast, if a better assignment can be found, we proceed with the RB re-allocation. First, for the case of MUEs (line 7), such a re-allocation is manipulated in the same way as that in [13], by reassigning RBs of CC c_1^* (or c_2^* if specified) from the DeNB that can contribute more utility to MUE u_1^* (or u_2^*) (lines 8–14). However, the current utility values contributed by these RBs perceived by not only the DeNB but also all RNs need to be updated (lines 10 and 14).

On the other hand, if RBs of CC c_1^* are reassigned to UE u_1^* via RN a^*, we only reassign those RBs selected by Subroutine_1 to u_1^* (line 16). Then we recalculate the current total utility value contributed by u_1^* on RBs from the DeNB via a^* in line 17 and update those contributed by the foregoing RBs perceived by all base stations in line 18. Moreover, we re-allocate those RBs of CC c_2^* obtained in Subroutine_1 from RN a^* to u_1^*, with the MCS also determined by the same subroutine, for realizing the bandwidth of CC c_1^* newly allocated from the DeNB (line 19). Accordingly, we recalculate the present total utility value accomplished by u_1^* on RBs from RN a^* in line 20. Remark that in line 21, we also set the utility values of those above RBs of c_2^* perceived by all base stations nearby, including a^* and the DeNB, to infinite so as to

(01) Calculate $h_D(u,c,b,\ell,a), \forall u \in \mathscr{S}_a, \forall c \in \mathscr{C}, \forall b \in \mathscr{B}, \forall \ell \in \mathscr{L}, \forall a \in \mathscr{A}$:
(02) Calculate $h_R(u,c,b,\ell,a), \forall u \in \mathscr{S}_a, \forall c \in \mathscr{C}, \forall b \in \mathscr{B}, \forall \ell \in \mathscr{L}, \forall a \in \mathscr{R}$;
(03) **repeat**
(04)   Calculate $g(u_1,u_2,c_1,c_2,\ell_1,\ell_2,a), \forall (u_1,u_2,c_1,c_2,a) \in \mathscr{M}, \forall \ell_1 \in \mathscr{L}, \forall \ell_2 \in \mathscr{L} \cup \{0\}$
           by Subroutine_1;
(05) $(u_1^*,u_2^*,c_1^*,c_2^*,\ell_1^*,\ell_2^*,a^*) \leftarrow argmax_{\forall (u_1,u_2,c_1,c_2,a) \in \mathscr{M}, \forall \ell_1 \in \mathscr{L}, \forall \ell_2 \in \mathscr{L} \cup \{0\}} g(u_1,u_2,c_1,c_2,\ell_1,\ell_2,a)$;
(06)   **if** $g(u_1^*,u_2^*,c_1^*,c_2^*,\ell_1^*,\ell_2^*,a^*) = 0$ **then goto** line 39;
(07)   **if** $a^* = 0$ **then**
(08)   Assign CC c_1^* from the DeNB to UE u_1^*;
(09)   **for each** RB $b \in \mathscr{B}$ with $h_D(u_1^*,c_1^*,b,\ell_1^*,0) > \mathscr{H}(c_1^*,b,0)$ **do**
(10)   Reassign b of c_1^* with MCS ℓ_1^* to UE u_1^*; $\mathscr{H}(c_1^*,b,a) \leftarrow h_D(u_1^*,c_1^*,b,\ell_1^*,0), \forall a \in \mathscr{A}$;
(11)   **if** $u_2^* \neq 0$ **then**
(12)   Assign CC c_2^* from the DeNB to UE u_2^*;
(13)   **for each** RB $b \in \mathscr{B}$ with $h_D(u_2^*,c_2^*,b,\ell_2^*,0) > \mathscr{H}(c_2^*,b,0)$ **do**
(14)   Reassign b of c_2^* with MCS ℓ_2^* to UE u_2^*; $\mathscr{H}(c_2^*,b,a) \leftarrow h_D(u_2^*,c_2^*,b,\ell_2^*,0), \forall a \in \mathscr{A}$;
(15)   **else**
(16)   Reassign all RB(s) in $\mathscr{D}^*$, obtained in Subroutine_1, with MCS ℓ_1^* to RN a^* for UE u_1^*;
(17) $\mathscr{K}_D(u_1^*) = \mathscr{K}_D(u_1^*) + \sum_{\forall b \in \mathscr{D}^*} h_D(u_1^*,c_1^*,b,\ell_1^*,a^*)$;
(18) $\mathscr{H}(c_1^*,b,a) \leftarrow h_D(u_1^*,c_1^*,b,\ell_1^*,a^*), \forall b \in \mathscr{D}^*, \forall a \in \mathscr{A}$;
(19)   Reassign all RB(s) in $\mathscr{E}^*$ with MCS ℓ^* from RN a^*, obtained in Subroutine_1, to UE u_1^*;
(20) $\mathscr{K}_R(u_1^*) = \mathscr{K}_R(u_1^*) + \sum_{\forall b \in \mathscr{E}^*} h_R(u_1^*,c_2^*,b,\ell^*,a^*)$;
(21) $\mathscr{H}(c_2^*,b,a) \leftarrow \infty, \forall b \in \mathscr{E}^*, \forall a \in \mathscr{N}_{a^*} \cup \{0,a^*\}$;
(22)   **for each** RB b of CC c_2^* in $\mathscr{E}^*$ used by DeNB **do** DeNB abandons using b of c_2^*;
(23)   **for each** $b \in \mathscr{E}^*$ and each $a \notin \mathscr{N}_{a^*} \cup \{0,a^*\}$ **do**
(24)   **if** $\mathscr{H}(c_2^*,b,a) \neq \infty$ **then** $\mathscr{H}(c_2^*,b,a) \leftarrow 0$;
(25)   Sort all pairs (c,b) in $\mathscr{V}_R(u_1^*)$ in the descending order of $\ell_{u_1^*,c}^R$ and
           arranging all pairs (c,b) in adjacent entries for each CC $c \in \mathscr{C}$;
(26)   Release RBs of RNs that accomplish the bandwidth of c_1^* from DeNB by Subroutine_2;
(27)   **if** $c_2^* \neq 0$ **then**
(28)   Release RBs of RNs that accomplish the bandwidth of c_2^* from DeNB by Subroutine_2;
(29)   Remove $(u_1^*,u_2^*,c_1^*,c_2^*,a^*)$ from $\mathscr{M}$;
(30)   Remove $(u_1^*,u,c,c',a), \forall u \in \mathscr{S}_0 \cup \{0\}, \forall c \in \mathscr{C}, \forall c' \in \mathscr{C} \cup \{0\}, \forall a \in \mathscr{Z}_{u_1^*} - \{a^*\}$, from $\mathscr{M}$;
(31)   **if** $u_2^* \neq 0$ **then**
(32)   Remove $(u_1^*,0,c_1^*,0,0)$, $(u_2^*,0,c_2^*,0,0)$ and $(u_2^*,0,c,c',a), \forall c \in \mathscr{C}, \forall c' \in \mathscr{C}$,
           $\forall a \in \mathscr{Z}_{u_2^*} - \{0\}$, from $\mathscr{M}$;
(33)   **if** $a^* \neq 0$ **then**
(34)   Remove $(u,0,c_2^*,c,a^*)$ and $(u,0,c,c_1^*,a^*), \forall u \in \mathscr{S}_{a^*}.\forall c \in \mathscr{C}$, from $\mathscr{M}$;
(35)   **if** RN a^* (or UE u_1^*) has employed f_{a^*} (or $k_{u_1^*}$) CCs **then**
(36)   Remove tuples corresponding to a^* (or u_1^*) and any CC unused by a^* (or u_1^*) from $\mathscr{M}$;
(37)   **else if** UE u_1^* (or u_2^*) is assigned $k_{u_1^*}$ (or $k_{u_2^*}$) CCs from DeNB **then**
(38)   Remove all tuples corresponding to UE u_1^* (or u_2^*) from $\mathscr{M}$;
(39) **until** $\mathscr{M} = \emptyset \vee g(u_1^*,u_2^*,c_1^*,c_2^*,\ell_1^*,\ell_2^*,a^*) = 0$;

Fig. 5 Our greedy scheduling algorithm

prohibit these stations from employing the same RB due to the interference control. Thus the DeNB needs to abandon using any such RB of CC c_2^* if it is currently using the RB (line 22). Now that the DeNB does not employ any foregoing RB of c_2^*, an RN that is not adjacent to a^* is allowed to use one without any penalty by setting the corresponding utility perceived by it to zero if none of its neighboring RNs is currently using such an RB (lines 23–24). Furthermore, in line 25, we sort all pairs (c, b) of $\mathscr{V}_R(u_1^*)$, i.e., CC-RB pairs assigned from RN a^* to UE u_1^*, in the descending order of their associated MCSs as well as arranging all pairs with the same CC in adjacent entries. In doing so, when we release redundant RBs from some RN for a UE, we can first release those with lower MCSs such that more RBs may be freed for future reuse.

Then, we proceed with releasing redundant RBs originally employed by RNs for accomplishing the bandwidth of CC c_1^* from the DeNB by Subroutine_2 in line 26. Still, if the second CC is specified, i.e., some RN or the second UE is considered, we also need to free redundant RBs of RNs that originally realize the bandwidth of CC c_2^* from the DeNB by the same subroutine (lines 27–28). After finishing the assignment specified, some tuples need to be removed from $\mathscr{M}$ to make the scheduling algorithm able to be terminated in a polynomial time. The tuple corresponding to the assignment in question is first removed from $\mathscr{M}$ (line 29). Furthermore, since UE u_1^* is served by RN a^* or the DeNB, it cannot be served by any other RN such that all associated tuples must be removed from $\mathscr{M}$ (line 30). Similarly, if the second MUE is involved (line 31), because we assign any CC from the DeNB to an MUE simply once, CCs c_1^* and c_2^* cannot be assigned again to MUEs u_1^* and u_2^*, respectively, and MUE u_2^* can be merely served by the DeNB as well (line 32). On the other hand, if an RN is involved at this assignment (line 33), seeing that an outband RN cannot use the same CC for both transmission and reception, we discard all corresponding tuples from $\mathscr{M}$ in line 34. Moreover, if the number of CCs used by RN a^* (or UE u_1^*) has reached the limit, for any CC c unused by RN a^* (or UE u_1^*), all tuples associated with both a^* (or u_1^*) and c must be removed from $\mathscr{M}$ (lines 35–36). In contrast, if what in question is the DeNB and any of the two considered MUEs has been assigned with the maximum number of CCs, all tuples corresponding to such an MUE are removed from $\mathscr{M}$ (lines 37–38). At last, the main loop of our algorithm is terminated in line 39 when all potential assignments are processed or no better assignment can be found.

4.5   Complexity Analysis

Now we start analyzing the time complexity of the above greedy scheduling algorithm. Here, let the numbers of CCs, RBs and MCSs be i, j and p, respectively. First, we estimate the computational complexity of Subroutine_1, which is dominated by the loop from lines 11 to 21 with the complexity of $O(pj^2 \log j)$. Next, we consider the complexity of Subroutine_2. Since any UE can be merely served by a single base station, the most outer two loops of the foregoing subroutine totally iterate up to m times. Moreover, the complexity of such a subroutine is mainly determined by its inner loop from lines 6 to 10. Thus its complexity is $O(mn^2 ij)$.

At last, we calculate the time complexity of the whole algorithm. Its complexity is dominated by the following three actions: (i) Calculating the gains of all remaining potential assignments by Subroutine_1 in line 4; (ii) Sorting all CC-RB pairs used by the considered UE for realizing the bandwidth allocated from the DeNB in line 25; (iii) Releasing redundant RBs of RNs that originally accomplish the bandwidth from the DeNB by Subroutine_2 in lines 26 and 28. The computational complexities of the above three actions are individually $O(m^2 n p^3 i^2 j^2 \log j)$, $O(ij \log ij)$ and $O(mn^2 ij)$. Because the numbers of UEs and RBs are far larger than those of RNs and CCs, respectively, i.e., $m \gg n$ and $j \gg i$, the complexity of the entire algorithm is determined by the first action. Furthermore, since each possible assignment will be considered at most once and any UE will be served by up to one base station, the main loop of the algorithm (lines 3–39) will at most iterate mfk times, where f and k are the maximum numbers of CCs usable by any RN and UE, respectively. So the total time complexity of our first algorithm is $O(m^3 n f k p^3 i^2 j^2 \log j)$. Now that the number of CCs employable by an RN or a UE, i.e., f or k, is limited, it can be regarded as a constant. Also, the number of RNs is small. Thus the above complexity can be further reduced to $O(m^3 p^3 i^2 j^2 \log j)$. This means that our heuristic scheduling algorithm will terminate in a moderate polynomial time.

5 Conclusions

In this paper, we have discussed how to extend the interesting idea introduced in [13] to efficiently perform joint downlink radio resource allocation in an LTE-A outband relay network with CA configuration, additionally taking the issues caused by multiple base stations into consideration. To the best of our knowledge, such an algorithm is the first one to this end. Here, due to the timeliness, we only presented our greedy scheduling algorithm first, and in the future, we will conduct a simulation study to verify the performance of our algorithm and compare its efficiency and complexity with the conventional methods adopted by LTE-A relay networks. Moreover, the scheduling algorithm proposed in this paper does not alter the MCSs of RBs that have been allocated to UEs or RNs during the scheduling task performed at each TTI. To achieve better performance, we will also consider how to design a scheduling algorithm that can change MCSs of used RBs to further boost the UE utility contributed by those RBs.

Acknowledgements This work was supported by the Ministry of Science and Technology, Taiwan, ROC, under Grant MOST 106–2221–E–005–021.

References

1. 3rd Generation Partnership Project (3GPP): Evolved Universal Terrestrial Radio Access (E-UTRA) and Evolved Universal Terrestrial Radio Access Network (E-UTRAN): Overall Description: Stage 2. TS 36.300 (2012)
2. 3rd Generation Partnership Project (3GPP): Evolved Universal Terrestrial Radio Access (E-UTRA): Physical Channels and Modulation. TS 36.211 (2012)

3. 3rd Generation Partnership Project (3GPP): Evolved Universal Terrestrial Radio Access (E-UTRA): Physical Layer Procedures. TS 36.213 (2012)
4. 3rd Generation Partnership Project (3GPP): Feasibility Study for Further Advancements for E-UTRA (LTE-Advanced). TR 36.912 (2011)
5. Andrews, M.: A survey of scheduling theory in wireless data networks. IMA Vol. Math. Appl. **143**, 1–18 (2007)
6. Aydin, M.E., Kwan, R., Wu, J.: Multiuser scheduling on the LTE downlink with meta-heuristic approaches. Phys. Commun. **9**, 257–265 (2013)
7. Chen, X., Suh, Y.H., Kim, S.W., Youn, H.Y.: Handover management based on asymmetric nature of LTE-A HetNet using MCDM algorithm. Int. J. Netw. Distrib. Comput. **3**(4), 250–260 (2015)
8. Deghani, M., Arshad, K.: LTE-advanced radio access enhancements: a survey. Wirel. Pers. Commun. **80**(3), 891–921 (2014)
9. Guan, N., Zhou, Y., Tian, L., Sun, G., Shi, J.: QoS guaranteed resource block allocation algorithm for LTE systems. In: Proceedings of IEEE 7th International Conference on Wireless and Mobile Computing, Networking and Communication (WiMob), pp. 307–312 (2011)
10. Ju, H., Liang, B., Li, J., Yang, X.: Dynamic joint resource optimization for LTE-advanced relay networks. IEEE Trans. Wirel. Commun. **12**(11), 5668–5678 (2013)
11. Kushner, H.J., Whiting, P.A.: Asymptotic properties of proportional-fair sharing algorithms. In: Proceedings of IEEE 10th Annual Allerton Conference on Communication, Control, and Computing, pp. 1051–1059 (2002)
12. Lai, W.K., Lin, M.T., Kuo, T.H.: Burst transmission and frame aggregation for VANET communications. Int. J. Netw. Distrib. Comput. **5**(4), 192–202 (2017)
13. Liao, H.S., Chen, P.Y., Chen, W.T.: An efficient downlink radio resource allocation with carrier aggregation in LTE-advanced networks. IEEE Trans. Mobile Comput. **13**(10), 2229–2239 (2014)
14. Salem, M., Adinoyi, A., Rahman, M., Yanikomeroglu, H., Falconer, D., Kim, Y.D.: Fairness-aware radio resource management in downlink OFDMA cellular relay networks. IEEE Trans. Wirel. Commun. **9**(5), 1628–1639 (2010)
15. Tian, H., Gao, S., Zhu, J., Chen, L.: Improved component carrier selection method for non-continuous carrier aggregation in LTE-advanced systems. In: Proceedings of IEEE Vehicular Technology Conference (VTC Fall), pp. 1–5 (2011)
16. Wang, Y., Pedersen, K.I., Sorensen, T.B., Mogensen, P.E.: Carrier load balancing and packet scheduling for multi-carrier systems. IEEE Trans. Wirel. Commun. **9**(5), 1780–1789 (2010)

Efficient Feature Points for Myanmar Traffic Sign Recognition

Kay Thinzar Phu and Lwin Lwin Oo

Abstract This paper proposes an efficient feature points for traffic sign recognition (TSR) system. This system composed of adaptive thresholding method based on RGB color detection, shape validation, feature extraction and adaptive neuro fuzzy inference system (ANFIS). There are some important issues for real-time TSR system such as; lighting conditions, faded color traffic signs and weather conditions. The proposed adaptive thresholding method overcomes various illumination and poor contrast color. Features play main role in TSR system. The significant feature points such as termination points, bifurcation points and crossing points are proposed. This proposes feature points provide good accuracy in TSR system. Lastly ANFIS is used to recognize the proposed feature points. This system showed that this proposed method can achieve cloudy and drizzle rain condition. In this system, this proposed method is used to evaluate on Myanmar Traffic Sign data.

Keywords Adaptive thresholding based on RGB color · Aspect ratio · Bifurcation point · Crossing point · Termination point

1 Introduction

Traffic signs are involved in main point to support safely on the road and reduce the number of traffic accidents and to remind valuable information for drivers about regulations, limitations, directions and status of the road etc. Drivers sometimes lack of knowledge or miss about the actual road signs that may cause traffic accidents and jammed. In order to solve the traffic problems and safety, automatic TSR system becomes a key part of an advanced driver assistance systems and intelligent trans-

K. T. Phu (✉)
University of Computer Studies, Mandalay, Myanmar
e-mail: kaythinzarphue@gmail.com

L. L. Oo
Department of Natural Science, University of Computer Studies, Mandalay, Myanmar
e-mail: lwinlwin0072@gmail.com

© Springer Nature Switzerland AG 2019
R. Lee (ed.), *Computer and Information Science*, Studies in Computational
Intelligence 791, https://doi.org/10.1007/978-3-319-98693-7_10

portation system. As technologies offer more and more processing power, the goal of real time traffic sign detection and recognition is becoming feasible. Some new models of high class cars already well equipped with ADA system.

To develop and automatic TSR system has big challenges: illumination changes, lighting condition differs according to the time of the day, season, cloudiness and other weather condition, etc. Blurring effect and fading of traffic signs are other challenging factors affecting the traffic sign detection. The aim of this system is to implement efficient and effective TSR system for real-time. This ongoing work, RGB color based adaptive thresholding method is proposed for red color, yellow color and blue color traffic signs detection from video. Aspect ratio method is used to confirm traffic sign. New features points are extracted and given as input to ANFIS. The proposed system was to overcome the current challenges of traffic signs such as illumination changes.

The researchers have already proposed various ways to minimize the problem. Nevertheless, the recognition of narrow-gauge road signs is announced in real time. Today, many of traffic accidents have happened with loss of many human resources. TSDR system assists the driver to control their car in-time by alerting the upcoming road signs and potentially risky of road. There still exist variations of traffic signs between different countries that have signed agreements. The variations seem to be no importance to a human, but could pose significant challenges for a computer vision algorithm [1]. In Myanmar, there are three types of traffic signs: red, blue, and yellow color and shapes are circle and diamond. First, prohibitory signs are circle sign with white background color, and inner pictogram is black surrounded by red color. Moreover, warning signs are yellow diamond shape with black color border and inner pictogram is black, and then, ordered signs are blue circle with white inner pictogram. In this paper, we focus on blue, red and yellow color traffic signs for real-time.

This paper is comprised as follows. Section 2 describes the literature review of TSR system. Section 3 presents the details of the proposed method. Sections 4 and 5 display the experimental results and conclusion.

2 Literature Review

Until now, the research in Traffic Sign Recognition system has been centered on European traffic signs, but signs can look very different across different parts of the world. The author proposed RGB color model is converted to the HSL color model and color filter is used to extract the traffic signs from candidate image. For shape analysis, Harris corner detector is used on Myanmar traffic signs in [2]. Reference [3] presented grayscale image is converted to binary image. They used circle finding algorithm to detect only circular object. The authors proposed six features which will be given as input to the ANFIS for recognition. They obtained a satisfactory result which leads to the use of ANFIS system for traffic sign recognition. Møgelmose proposed detection methods on US traffic sign (focused on speed-limit sign) detection:

Integral Channel Features and Aggregate Channel Features in [4]. Reference [5] presented red color enhancement and thresholding in preprocessing step. In detection step, filter and detect MSER regions in traffic signs images. The paper proposed an RGB-based color thresholding technique with fixed threshold and developed circle detection algorithm [6]. The authors enhanced through Weiner filter and histogram equalizer. The region of interests is detected using the improved Connected Components Labeling algorithm combined with Vectorization to reduce computational time in [7]. The researcher performed histogram equalization technique to enhance the image and media filter to remove noise. The image is converted to binary image by threshold value 0.18 and center, width and height are calculated for shape validation [8]. The author used RGB color thresholding and morphological closing operator for segmented. Circular Hough transform is employed distinction between traffic signs and other objects [9]. RGB histogram equalization, RGB color segmentation, modified grey scale segmentation, binary image segmentation and shape matching are used for Malaysia standard and non-standard road signs detection in [10].

In traffic sign recognition stage, converted color image to grayscale. And then, Sobel edge detector is used to segment road sign. Lastly, the segmented road signs are matching with images stored in the database by using template matching method in [2]. The SURF descriptor is used to produce Bag of words and k-means clustering is used to get the discriminative features. Artificial Neural Network is used for recognition [8]. Reference [6] combined Gabor features, local binary patterns, and histogram of oriented gradients within a support vector machine classification framework. The histogram matching are used the detected image with other images stored in the database [7]. References [5, 9] also used HOG features and SVM for recognition. Islam [10] proposed 278 features points for each road signs and ANN is used for recognition.

Reference [3] was detected red circular traffic signs and Malaysian traffic signs dataset in [8, 10]. Circular traffic signs images from GTSDB dataset are used in [6], and the researchers of [7, 8] also can detect for red traffic signs and speed limit signs only.

3 Proposed Method

The main architecture of proposed system is in Fig. 1. This system presents a positioning area of the traffic sign, the detection step, the characteristic extraction step and the identification step. Firstly, the area of interest as a traffic sign is located from the frame of the input image. In detection step: color detection process, board extraction and lastly its shape recognition process. The color of interest as a traffic sign is detected from the localized input image in color detection step. The result of the color detection process includes a binary image. And then, the detected region is extracted as board. The second process applies a process of verification of form to the board image to determine if the traffic sign. The candidate symbols are verified, which is the input of the system feature extraction step. Finally, the system identifies

Fig. 1 Architecture of the
proposed system

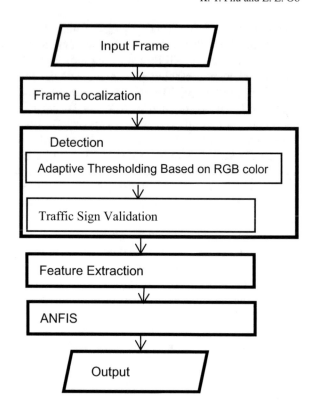

the extracted feature points and provides the results. This article will focus on the
Myanmar red, yellow and blue road signs.

3.1 Traffic Sign Localization

This step is designed to detect the exact location of the traffic sign in the input video
frame. The system extracted each first frame in three frames. In Myanmar, the car
runs on the right side of the road and the traffic signs appear on the right side of the
frame of the image. Therefore, the region of interest (ROI) where the symbol appears
is extracted from the frame of the image saved in the fast loading time. Therefore,
the process rejects areas that do not include traffic signals. The system extracts the
fourth areas from the row of the column and the fourth area from the right half. The
following Fig. 2 shows the candidate areas using the proposed method.

| (a) Original Frame | (b) Localized Output frame |

Fig. 2 Localization of traffic sign

3.2 Traffic Sign Detection

In this step, system detects traffic sign in the localized frame. The traffic signs color is mixture with environment color e.g. blue color of traffic sign is wrong with sky color. Therefore, the color detection is used to select traffic sign color.

3.2.1 RGB Color Thresholding

There are various techniques by which adaptive form of thresholding selection can be utilized, but the aim of all the ways is to find the appropriate threshold for different regions in the image [11]. Adaptive thresholding is used to separate desirable foreground image objects from the background based on the difference in pixel intensities of each region. This method provides achieve results in uniform illumination. In this process, thresholding based on RGB color method detects traffic sign in input frame. Blue pixels values are extracted from red and green pixels values to get yellow color. After getting the maximum thresholding value, the actual traffic signs pixel value is less than maximum thresholding value. By experiments, the maximum thresholding value is reduced 35% to detect real yellow color traffic sign board. The following equations are used to detect yellow color traffic signs.

$$TS_{rb} = \max(R_{(i,j)} - B_{(i,j)}) * 0.65$$
$$TS_{gb} = \max(G_{(i,j)} - B_{(i,j)}) * 0.65 \qquad (1)$$

$$Yellow_{(i,j)} = \begin{cases} YI_{(i,j)} = 1 & \begin{array}{l} \text{if}(\max(R_{(i,j)} - B_{(i,j)}) \geq TS_{rb})^{\wedge} \\ (\max(G_{(i,j)} - B_{(i,j)}) \geq TS_{gb}) \end{array} \\ YI_{(i,j)} = 0 & \text{else} \end{cases} \qquad (2)$$

Blue pixel values are greater than red and green pixel value in blue color traffic sign. So, we calculate blue color traffic signs using the following equations.

$$TSB_{rb} = \max(B_{(i,j)} - R_{(i,j)}) * 0.65$$
$$TSB_{gb} = \max(B_{(i,j)} - G_{(i,j)}) * 0.65 \tag{3}$$

$$Blue_{(i,j)} = \begin{cases} BI_{(i,j)} = 1 & \begin{aligned} &if(\max(B_{(i,j)} - R_{(i,j)}) \geq TSB_{rb})^{\wedge} \\ &(\max(B_{(i,j)} - G_{(i,j)}) \geq TSB_{gb}) \end{aligned} \\ BI_{(i,j)} = 0 & \text{else} \end{cases} \tag{4}$$

For the red color, the following equations are also used to find the red color traffic signs in the input image frame.

$$TSR_{rb} = \max(R_{(i,j)} - B_{(i,j)}) * 0.65$$
$$TSR_{gb} = \max(R_{(i,j)} - G_{(i,j)}) * 0.65 \tag{5}$$

$$Red_{(i,j)} = \begin{cases} RI_{(i,j)} = 1 & \begin{aligned} &if(\max(R_{(i,j)} - B_{(i,j)}) \geq TSR_{rb})^{\wedge} \\ &(\max(R_{(i,j)} - G_{(i,j)}) \geq TSR_{gb}) \end{aligned} \\ RI_{(i,j)} = 0 & \text{else} \end{cases} \tag{6}$$

where R is red, B is blue and G is green color intensities value with range from 0 to 255. TSrb, TSgb are red threshold value and blue threshold values are TSBrb, TSBgb and TSRrb, TSRgb are the threshold values for red.

3.2.2 Shape Validation

There are images of the same color and shape as traffic sign. It is required to validate the traffic sign from output binary image. The removal of non-traffic sign regions from detected image involves two-steps. In the first step, the holes of output binary image are filled and then labeled the 8 connected neighbors pixels are grouping as one potential region. Then, bounding box characteristic (height and width) of row and column is calculated and to define a set of potential region R = {R1, R2, ..., RN} where N is the number of potential TS regions. Each region contains 4 values (maximum and minimum pixel values of row and column). In our implementation, total number of region values are more than 10 regions, eliminate this region. We choose this normalized value based on various experiments and a statistical study. In the second step, aspect ratio method is used to validate the detected image is traffic sign or not. The following equations are used to calculate the shape of the candidate region.

$$AspRatio = \frac{BDWidth}{BDHeight}$$

$$AbsRatio = \begin{cases} 1 \ if \ (1 - AspRatio) < 0.25 \\ 0 \ else \end{cases} \tag{7}$$

where, AspRatio is aspect ration, AbsRatio is absolute aspect ratio, BDWidth and BDHeight are width and height of the color detected board image.

3.3 Feature Extraction

The main contribution of this system is the extraction of feature points. The TSDR system needs to identify the output image of the detection phase. The traffic sign detection procedure has several characteristics. Need to extract important features of the TSDR system. The main characteristic of the TSDR system is its fast processing speed. In the feature extraction process, the output of the shape verification traffic signs image uses the morphological operator to thin the inner pictogram. Second, small regions are refine by using morphological opening. The four main characteristics: the end point, the branch point and the crossing points are extracted from the refined image. The TSDR system obtains accurate results by using the suggested feature points. The following formula is used to find the end point in the candidate area.

3.3.1 End Point

The following equation is used to find termination points in validated traffic sign.

$$Tr = \begin{cases} 1 \text{ if}((Cen == 1)^{\wedge} sum(N == 1)) \\ 0 \text{ else} \end{cases} \tag{8}$$

3.3.2 Branch Point

The following equation is used to find the bifurcation points from output sign board.

$$Bif = \begin{cases} 1 \text{ if}((Cen == 1)^{\wedge} sum \\ 0 \text{ else} \end{cases} \tag{9}$$

where, Tr = Termination point, Bif = Branch point, Cen = Center point in 3×3 matrix, N = 3×3 matrix.

The following figures show the termination points and bifurcation points that are extracted for T turn yellow color traffic sign (Fig. 3).

Some traffic signs produce the same number of features points (for example, the Y junction produces 3 end points, and a fork point and the T junction point also produce the same characteristic points). Therefore, this system calculates the average of the end points and the bifurcation points of the rows and columns. Therefore, the four

Fig. 3 End and branch
points

(a) Original image   (b) 1 branch and 3
                                          end points

characteristic points of end and branch are extracted. The following Eqs. (10) and
(11) are used to calculate average values of termination and bifurcation points.

$$Avg\_Tr\_R = \frac{mean\,(Tr)}{Rr}$$

$$Avg\_Tr\_C = \frac{mean\,(Tr)}{Cc} \tag{10}$$

$$Avg\_Bf\_R = \frac{mean\,(Bf)}{Rr}$$

$$Avg\_Tr\_C = \frac{mean\,(Bf)}{Cc} \tag{11}$$

where, Avg_Tr_R and Avg_Tr_C are row and column average values of termination
points, Rr and Cc are total row and column, Avg_Bf_R and Avg_Bf_C are row and
column average values of bifurcation points.

3.3.3 Crossing Point

The number of transitions from the background to the foreground pixel along the vertical and horizontal lines through the image is calculated by crossing [12]. Figure 4a
below shows the horizontal and vertical intersections. In this system, the detected
traffic signs are divided into four rows: one third and two thirds of the rows and
columns. The average position value of the intersections in each dividing line is calculated for the characteristic points. Figure 4b below shows the four intersections of
the traffic signs produced by the system. Table 1 lists the discussion image for eight
functions.

$$VC\_1 = \frac{mean\,(C1)}{Cc}$$

$$VC\_2 = \frac{mean\,(C2)}{Cc} \tag{12}$$

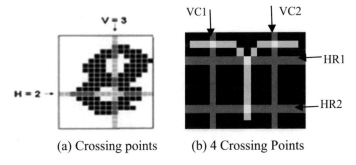

(a) Crossing points   (b) 4 Crossing Points

Fig. 4 Crossing feature points

Table 1 Proposed features points values for T turn

Name	Values
VC_1	0.166667
VC_2	0.166667
HR_1	0.5
HR_2	0.5
Avg_Tr_C	0.416666
Avg_Tr_R	0.5
Avg_Bf_C	0.272727
Avg_Bf_R	0.5

$$VR\_1 = \frac{\text{mean}(R1)}{Rr}$$
$$VR\_2 = \frac{\text{mean}(R2)}{Rr} \qquad (13)$$

where, C1, C2, R1 and R2 are total number of crossing points and VC_1, VC_2, VR_1 and VR_2 are average values of crossing points of one third and two third regions of row and column respectively.

3.4 Recognition

The ANFIS classifiers are used to recognize road traffic signs. ANFIS is with the training ability of a neural network and the advantage of a rule-based system. This is the main advantage of fast convergence. In this system, the ANFIS system is built with 8 inputs corresponding to 8 feature points and functions of the trapezium function and two membership functions used for input functions. The output data presented to each input was represented by the output label of each set of input and the hidden level varies according to the number of rules that provide the maximum recognition rate for each set of characteristics. Lastly, the rule output is combined into one conclusion functions (for example, 1 is designated as a Y junction indicator, 2 is a left turn indicator).

4 Experimental Results

In this system, the data is collected by using smartphones on highways. The image resolution in the video sequence is 720×480. The frame rate of the video is 30 frames/s. The speed of the vehicle is about 40 km/h. In this experiment, the cloudy days, drizzle rain and normal weather conditions videos are tested. In our experiment, this proposed method can detect traffic signs of more than 60 feet. Therefore, drivers will reduce and control their speed on time. The extracted feature points are used to identify real-time detected traffic signs. Figures 5, 6 and 7 of (a) show the original image extracted from the video file (b) shows the binary traffic sign board obtained using the proposed threshold method, (c) the image board is in the extracted color. In Fig. 5a, the traffic sign is captured in cloudy conditions, so the frame of the image is obviously not. The proposed method can be detected in such an image. The output binary image is shown in Fig. 5b. In Fig. 6, the image frame is covered by drizzle. This proposed detection method detects this situation. Figure 8 shows the system's identification results.

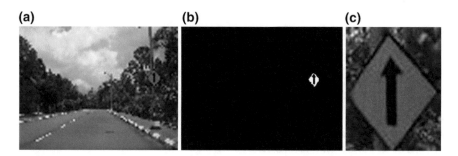

Fig. 5 Yellow color traffic sign under cloudy weather

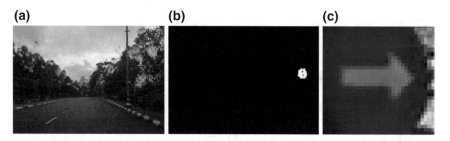

Fig. 6 Blue color traffic sign under drizzle rain weather

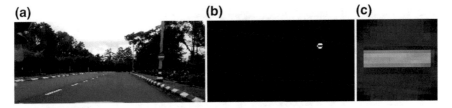

Fig. 7 Red color traffic sign at daytime

Fig. 8 Recognition result of this proposed system, the red box showed the output recognized traffic sign

The system tested several threshold ranges. As the threshold range begins to decrease, the number of detected frames increased, and the detected area does not contain traffic signs, which consumes the performance of the system. The recommended thresholds values of yellow, blue and red applied to all weather conditions, such as light rain and cloudy weather. The system tested different row and column locations to locate the area of the traffic sign in the input frame. If we increased the position of the rows and columns, the same color as the traffic sign are detected and a blurred image is produced. At the same time, the position of the rows and columns was also reduced, and the system omitted some areas of traffic signs. In this system, several durations of 16 videos were used for a total of 18 traffic signs, and

Table 2 Recognition accuracy by using proposed feature points

Traffic signs	Test images	Correctly classified signs	Accuracy (%)
Mandatory	152	141	93
Informative	145	138	95
Prohibition	122	120	98
Total	419	399	95

Table 3 Detection results by using proposed method

Myanmar traffic signs	Test images	Correctly detected signs	Accuracy (%)
Prohibitive	135	135	100
Mandatory	304	296	97
Informative	233	230	99
Total	672	661	98

Table 4 Compared result of proposed method and Win [2]

Method	Win [2]		Our method	
	Total tested signs		Total tested signs	
Detection rate	95%	83	98%	661
Recognition rate	94%		95%	399

the accuracy was calculated for this system. The results are shown in Tables 2 and 3 and comparison results are showed in Table 4. The system produces recognized traffic signs that are exported in Myanmar language.

5 Conclusion

The TSDR system is an important part of the driver assistance system. It warns and guides the drivers about the road conditions. These systems must quickly and accurately capture the traffic signs of the video sequences for real-time. The new detection method based on RGB color values is proposed for detection and efficient feature points are extracted for traffic signs recognition in real time. The system is applied to the traffic signs data in Myanmar for red, yellow and blue color. The video was filmed in the morning, cloudy and rainy conditions. As mentioned above, the system tests different thresholds and localized locations. The suggested threshold and local values provide better performance for this system. The system provides effective detection and identification precision. In future studies we will keep some important feature points and other types of Myanmar traffic signs.

References

1. Deshpande, A.V.: A brief overview of traffic sign detection methods. Int. J. Eng. Res. 141–144 (2016)
2. Win, W.S., Myint, T.: Detection and recognition of Myanmar traffic signs from video. In: Proceedings of 2015 International Conference on Future Computational Technologies, (ICFCT'2015), Singapore. ISBN 978-93-84468-20-0
3. Billah, M., Waheed, S., Ahmed, K., Hanifa, A.: Real time traffic sign detection and recognition using adaptive neuro fuzzy inference system. Commun. Appl. Electron. (CAE) 3(1) (2015). ISSN 2394 – 4714
4. Mogelmose, A., Liu, D., Trivedi, M.M.: Detection of US traffic signs. IEEE Trans. Intell. Transp. Syst. (2015)
5. Agrawal, S., Chaurasiya, R.K.: Ensemble of SVM for accurate traffic sign detection and recognition. In: ICGSP, Singapore (2017)
6. Berkaya, S.K., Gunduz, H., Ozsen, O., Akinlar, C., Akinlar, S.: On circular traffic sign detection and recognition. Expert Syst. Appl. 48, 67–75 (2016)
7. Daraghmi, Y.A., Hasasneh, A.M.: Accurate real-time traffic sign recognition based on the connected component labeling and the color histogram algorithms. In: International Conference on Signal Processing (2015)
8. Aparna, S., Abraham, D.: Multiple traffic sign detection and recognition using SVM. Int. J. Adv. Res. Comput. Sci. Manag. Stud. 3(8) (2015)
9. Islam, Kh.T., Raj, R.G.: Recognition of traffic sign based on bag-of-words and artificial neural network. Symmetry 138 (2017)
10. Islam, Kh.T., Raj, R.G.: "Real-time (vision-based) road sign recognition using an artificial neural network", Sensors, 17(4), p. 853, 2017
11. Sharma, P., Abrol, P.: Color based image segmentation using adaptive thresholding. Int. J. Sci. Tech. Adv. 2(3), 151–156 (2016)
12. Kumar, G., Bhatia, P.K.: A detailed review of feature extraction in image processing systems. In: Advanced Computing & Communication Technologies (ACCT), Feb 2014

A Study on the Operation of National Defense Strategic Operation System Using Drones

Won-Seok Song, Si-Young Lee, Bum-Taek Lim, Eun-tack Im and Gwang Yong Gim

Abstract As electronic warfare (EW) fused with the 4th Industrial Revolution such as unmanned technology, cyber warfare and AI (Artificial Intelligence) has played increasing roles in NCOE (Network Centric Operation Environment), which is a future battlefield environment, it is more important than ever to detect and identify an enemy's attack, and cope with electromagnetic waves. In addition, it seems important to calculate weight to use when we define the operational concept of weapons system and the elements suitable for the system and select the sectors where they can be highly useful in order to elevate the operational effectiveness. Unfortunately, however, it has been known that there is no case where the opinions of an expert group are given to the application of EW system or weight for each army by EW sector is applied, though the system has been actively developed through domestic and overseas R&D. With the issue in mind, I will use 'Electronic Warfare Subdivisions' (Fig. 1) of Joint Chiefs of Staff (hereinafter referred to as "JCS") in reviewing priority in preparation for EW system to mount on drones and examine the difference in priority by EW sector. In this paper, I will explain the concept and prospect of drones and EW system and propose a method of applying a relative weight and calculating importance (priority) by EW sector in order to enhance the capability of future EW.

Keywords EW (Electronic Warfare) · (Drones) · UAV (Unmanned Aerial Vehicle) · SONATA(SLQ-200(V)K)

W.-S. Song · S.-Y. Lee · B.-T. Lim
Department of IT Policy and Management, Soongsil University, Seoul, Republic of Korea
e-mail: sws.itpm12@gmail.com

S.-Y. Lee
e-mail: sylee@datasolution.kr

B.-T. Lim
e-mail: neobtlim@gmail.com

E. Im · G. Y. Gim (✉)
Department of Business Administration, Soongsil University, Seoul, Republic of Korea
e-mail: gygim@ssu.ac.kr

E. Im
e-mail: iet030507@gmail.com

© Springer Nature Switzerland AG 2019
R. Lee (ed.), *Computer and Information Science*, Studies in Computational Intelligence 791, https://doi.org/10.1007/978-3-319-98693-7_11

1 Introduction

EW must be a very strange term to the general public. However, the media paid attention to it once when mobile phone users, airplanes and ships including warships complained of the difficulty in receiving GPS signals due to radio-frequency disturbance (jamming) caused by the North Korea's attack to GPS (Global Positioning System). North Korea attempted jamming by using the modern EW equipment that it developed by itself or the communication and radar jamming devices that it imported from Russia in the past. This act can be defined as electronic attack (EA).

Like this, EW is the act to control, attack or disturb an enemy's electromagnetic spectrum or directed energy weapons with core military strengths of high technology. EW can be sub-categorized into electronic warfare support (ES), electronic protection (EA), and electronic protection (EP) as seen in Fig. 1.

Electronic warfare support (ES) is the act of blocking the electromagnetic spectrum of our army and collect, monitor, and analyze the electromagnetic spectrum energy of an enemy to recognize its threat, and support EW activities (threat avoidance, tracking, and etc.). In using ECM (Electronic Counter Measures), EA means to attack manpower, facilities, and equipment to neutralize the weapons of electromagnetic radiations and eventually combat capability by having a direct impact on their electromagnetic energy. The examples of EA include communication jamming, radar disturbance, directed energy weapons/laser attack, expendables (flares and active decoys) and RCIED (Radio Controlled Improvised Explosive Device). Electronic protection (EP) generally means all of the activities taken to protect the

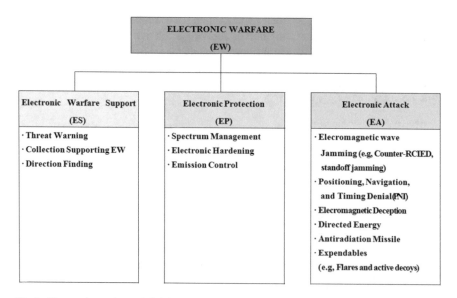

Fig. 1 Electronic warfare subdivisions [1]

military units, equipment, and an operational objective of our army from the enemy, and used to avoid the influence of our army's EA [1–6].

EW system is a weapons system that collects and analyzes the electromagnetic waves of the enemy's weapons system and identifies the location and operation conditions of their missiles, radar, and communication equipment. Based on the information, it radiates jamming electromagnetic waves to disturb the operation of the enemy's weapons system [7]. The main objective of EW system is to serve ES, EA, and EP to defend the warships of our army from the antiship missiles from the enemy's combat planes and warships [8].

For drones, which were first used in military sectors, many countries consider it as one of the future strategic industries. Private sectors have also carried out R&D in drone-related areas and applied drones to a variety of fields such as communication, aerial photograph, the supervision of crime, disaster, and catastrophe, the management of infrastructure, instrumentation, and etc.

To serve the objective of this study, Analytic Hierarchy Process (AHP) Model for calculating the factors of EW system and their weights so that EW can exert expected effectiveness when mounted on a drone in preparation for future combat operation competence was employed. In addition, Electronic Warfare Subdivisions of JCS [9] were subdivided to derive the persuasive factors for a decision hierarchy. A questionnaire was also constructed that consists of 3 top factors (ES, EA, EP) and 9 sub-questions in an attempt to calculate weights. The survey was carried out for the experts with 10-year or longer service period in EW field of the navy. Using the results of the survey, geometric mean was used to draw out weight and confirmed the reliability of the questionnaire by measuring the consistency of the experts' responses. The relative weights, which are the goal of this study, were obtained through this procedure.

Last, this dissertation is composed as follows: Sect. 1 Introduction where the necessity of a research in EW system, Sect. 2 Literature Review, Sect. 3 Research Method and Analysis, Sect. 4 Analysis Results, and Sect. 5 Conclusion. It is expected that when the national defense operation system is established, the results of this study will be used as reference data for prioritizing the sectors where drones can produce the most desirable effect with EW system mounted.

2 Literature Review

2.1 Concept and Prospect of EW System (SLQ-200K, SONATA)

A representative EW system that Korea developed with domestic technologies is 360° electromagnetic wave reception system. Installed in naval vessels and sites, the reception system quickly detects and analyzes the threat of an enemy's radar and guided missiles, and electromagnetic wave signals on the sea, and identifies

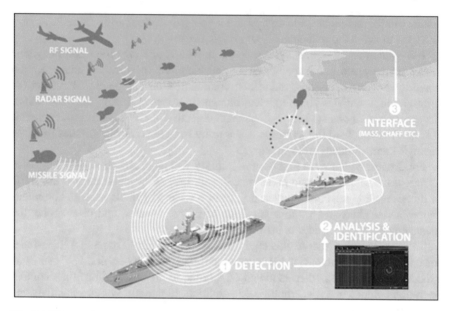

Fig. 2 EW operational concept

the threatening radio frequency signals and gives a warning signal. In addition, the system works with a guided missile defense system (e.g. R-BOC) mounted on a naval vessel to give off an alarm. Therefore, a quick response is possible. There is a demerit, however: if some of the signal information each site collects are ones already saved in EW system, threat cues should be prioritized and sent back to the collection units manually. Raw data, which are early detected source signals, should be analyzed to know the value of usability. The signals detected in a vessel site can be differently interpreted for their threat cues due to the mechanical error of EW system in bearings detection and depending on the operator's personal competence, all of which weaken the reliability of EW system. In particular, the strategic concept of EW, as seen in Fig. 2, is built on the basic concepts like a real-time reception of an enemy's threatening electromagnetic waves → detection, analysis, and identification → threat warning and recommendation of a countermeasure to cope effectively with the threats to naval vessels, airplanes, and missiles [6].

(1) Signal reception: receiving an enemy's threat signals (RF signal) in all directions
(2) Signal detection, comparison, analysis, and identification: the signal reception unit receives signals, detect bearings, and compares, analyzes, and identifies them
(3) Warning and responding: system setting and analysis, displaying of identified threat signals, warning, interlocking and recommending a countermeasure.

Compared with the advanced countries, Korea has kept focusing more on the protection of EW platform as a future business. In addition, as EW system has

Table 1 The domestic developments of EW system [6]

Model	Purpose
ULQ-11K/12K	Self-defense/escort protection
ALQ-88AK	F-4/F-16 fighter self-protection
APECS-II	Protecting the destroyer itself
SLQ-200K (SONATA)	Protecting the destroyer itself (Yoon Young Ha Class, Landing ship Tank)
ALQ-200K	KF-16 fighter protection itself
ALR-200K (Lynx ESM)	Helicopter-mounted signal detection
TLQ-200K (Next Generation EW)	Intercepting/interfering with vehicle-mounted communication signals
TAC-ELINT	RF-16 Aircraft carrying information collection

Table 2 The overseas developments of EW system [4]

Country	Model	Purpose
USA	NULKA (Decoy)	Trap self-protection
	EA-18G (Growler)	Trap self-protection
	RC-135 (Cobra Ball)	Aircraft electricity
	ASPJ (Imbedded)	Aircraft sign information collection
	AN/MLQ-40	Aircraft protection itself electronic interruption
Russia	Khibiny	Aircraft protection itself electronic interruption
	Krasukha-4	Ground station radar ES/EA
Italia	Nettuno	Trap self-protection
Israel	SEWS DV	Trap self-protection
China	NRJ6A/945P	Trap self-protection
	JH-7A	Aircraft electricity
	Y-8	Aircraft sign information collection
France	Dupuy de Lome	Electronic War
Germany	Oste	Electronic War

been changing smaller, lighter, and more integrated, it is evolved in a direction to the defense of multiple threats at a long distance. Table 1 shows the domestic developments of EW system.

In addition, the advanced countries such as America, Russia, and China are currently integrating warship-protection EW systems, which has been developed and used as an independent entity, to one single system. The integrated EW system includes various sensors mounted on the system to increase the detection efficiency and also countermeasures. Table 2 shows the overseas development of EW system.

Table 3 The development trends of the advanced countries [10]

Division	Pre-2010	2011–2020	Post-2020
USA	• Tactical information gathering equipment • Communication information collection equipment • Self-protection jammer • Remote support jammer • Development of information protection SW • Tactical Cyber Weapons	• Integrated signal information • Next-generation integrated electronic warfare equipment • Securing information literacy • Information intrusion/fault tolerance technology • Next Generation Cyber Weapons	• Next generation integrated signal information collection equipment • Next generation integrated electric field monitoring/electronic attack equipment • Securing the next generation information capability • Next Generation Cyber Weapons
Japan	• Tactical information gathering equipment • Self-protection jammer • Cyber troop operation • Development of automation system security system	• Integrated signal information collection equipment • Integrated electronic warfare equipment • Remote support jammer • Real-time intrusion detection/response • Hacking/virus production	• Next generation integrated signal information collection equipment • Next-generation integrated electronic warfare equipment • Creative Cyber Weapons • Automatic immune/auto restoration system
China	• Tactical information gathering equipment • Self-protection jammer • Cyber intrusion detection and response • Hacking/virus production	• Integrated signal information collection equipment • Integrated electronic warfare equipment • Remote support jammer • Information Strategy Technology • Creative Cyber Weapons	• Next generation integrated signal information collection equipment • Next-generation integrated electronic warfare equipment • Securing information literacy • Next Generation Cyber Weapons
Russia	• Signal information collecting equipment • Jammer for self/remote support • Development of information defense technology • Hacking/virus production	• Integrated signal information collection equipment	• Next generation integrated signal information collection equipment • Next generation integrated electric field monitoring/electronic attack equipment • Securing information literacy • Next Generation Cyber Weapons

The advanced countries have already shifted the development of EW system in the direction for sensor integration, network-centric EW, and the convergence of attack weapon sensors to improve combat capability. Table 3 shows the technological trends of EW system in the advanced countries.

2.2   The Concept and Prospect of Drones

A drone is a general term that refers to an airplane or a helicopter-shaped flight vehicle not controlled by a man inside it but wirelessly [11]. It is also called 'UAV' (Unmanned Aerial Vehicle), 'UAS' (Unmanned Aircraft System), or 'RPAS' (Remotely Piloted Aircraft System). Middle- or large-sized UAVs were often used for military purposes in the past. However, as small and cheap drones have recently been spreading for the purpose of hobby and leisure activities, they are domestically classified into 6 categories for unified naming: maximum takeoff weight, operational ceiling, driving method, takeoff and landing method, and kinetic energy (Fig. 3).

In addition, a drone can equip various equipment (e.g. optical and infrared devices and radar sensors) as seen in the figure, depending on a field where it is applied, and performs such missions as surveillance, reconnaissance, precision-precision guided munition, communication relay, EA/EP, decoy, and etc. Compared with existing manned aerial vehicle, UVA has no limit to missions to carry out, can fly for long hours, and performs all-around-clock and all-weather operation. Furthermore, it can mount various sensors for collecting images, video, and signal information for multiple purposes and locate a target and transmit combat situation in real time [13]. Although UAV includes both military and civilian unmanned aerial vehicles and I have to use the term UAV (Unmanned Aerial Vehicle) for a drone in this paper according to the notice of Korean Agency for Technology and Standards for UAV (Unmanned Aerial Vehicle) in the National Standard (KSW9000) Classification and Terminology. However, the term 'drone' is used instead for the sake of readers' better understanding (Fig. 4).

From early on, Israel directed all its strength toward developing drones that are cost effective. Focusing on developing the system for tactics surveillance and a small-scale attack in a limited area, Israel has succeeded in realizing a local target attack system and its operation concept. That is, the country possesses an outstanding technology

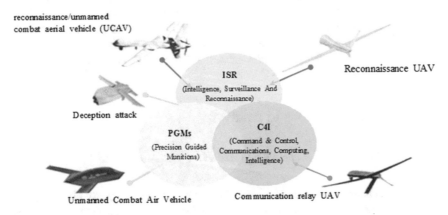

Fig. 3   The concept of future UAV operation [12]

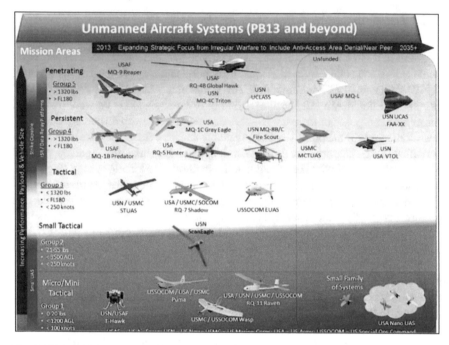

Fig. 4 UAS (PB13 & Beyond) [14]

related to the development of MALE (Medium Altitude Long Endurance) system or lower altitude rather than HALE (High Altitude Long Endurance) [4]. For China, Yilong, which is called 'the Chinese version of Predator', made its debut in 2013 Paris Air Show. In the show, it mounted BA-7 air-to-air missile (AAM), YZ-212 laser guided bomb (LGB), YZ-102 antipersonnel bomb, and 50 kg small guided bomb, and also displayed its performance as an unmanned attack plane. In addition, Xialong (the Chinese version of Global Hawk) is known to reach 57,000 feet high and fly up to 7,500 km far. It is considered that Xialong succeeded in a test flight in January 2013, flying 10 h at the speed of 750 km per hour, and is able to reconnoiter even the American territory of Guam, as well as South Korea and Japan [15].

3 Research Methods and Analysis

3.1 Application of AHP (Analytic Hierarchy Process)

AHP is a decision-making technique developed by Thomas L. Saaty (1980). Under various quantitative and qualitative criteria, it helps select optimal alternatives. It first hierarchizes a given decision-making problem. A pairwise comparison matrix a(n ×

n) is constructed with the relative importance or weight of each factor (or criterion) in the lower stratum right below the upper level. Weight is calculated with eigenvectors [16].

$$A\omega = \lambda_{max}\omega \tag{1}$$

(Here, A: pair wise comparison matrix, ω: eigenvector, λ_{max}: maximum eigenvector)

Enriched experience, intuition, and such of a decision maker are regarded as important. Therefore, qualitative information, which is difficult to measure but must be considered for decision making, as well as quantitative information can be handled fairly easily [1]. When expert knowledge is drawn out based on AHP, there are two important considerations: consistency ratio (CR) and integration of expert knowledge. CR is the tool to verify the reliability of expert knowledge. Thomas L. Saaty said that λ_{max-n} is a useful tool to measure consistency because it always has an equal value to or bigger value than n a positive transposed matrix. Only when matrix A is consistent, the value can be n [17]. Using this principle, we can get consistency index (CI).

$$CI = \mu = \lambda_{max-n/n-1} \tag{2}$$

(Here, λ_{max}: maximum eigenvector n: the umber factors)

When CI is divided by random index (RI), it yields consistency ratio (CR). Thomas L. Saaty maintained that when consistency is perfect, CR is equal to 0. On the contrary, when consistency is not perfect or poor, CR is bigger than 0. Therefore, when CR > 0.1, it is necessary to make a decision again or modify it [18]. However, studies of some social science fields accept it up to 0.2 (20%) for allowance because it is not easy to establish independence between the upper and lower criteria, considering the characteristics of questionnaire items [19].

Arithmetic mean and geometric mean are used to integrate expert opinions. Woo Chun-sik et al. said in their comparative study of bankruptcy prediction model [15] that geometric mean should be used for qualitative information to aggregate expert opinions and it has a higher prediction rate of bankruptcy than when arithmetic mean is used. Therefore, geometric mean can aggregate expert opinions more firmly [16]. To design a model to apply to the given decision-making problem, using AHP, 4 steps should be followed as seen in Fig. 5.

Step 1: Make a decision hierarchy with strata (or levels) (environmental scenarios, criteria, and alternatives to implement) for a decision-making problem set on the objective. Namely, it is to hierarchize the criteria for evaluation. The factors in a lower stratum should be described more specifically and in detail. However, too many attributes in one level produce too many for relative comparison. Therefore, it is desirable that the number of criteria in one level should be no more than 9 [20].

Fig. 5 The developmental
procedure of one AHP
Model [21]

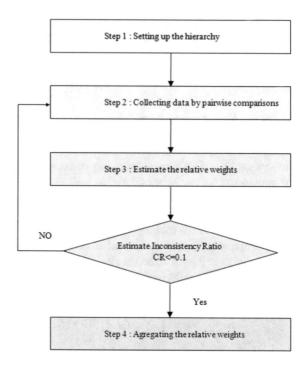

Step 2: Collect the data for the pairwise comparison of each criterion and its alternative in the environmental scenario. Make AHP matrix with the pairwise comparison (pairwise comparison matrix) data by level (stratum). That is, once a decision hierarchy is formed, make a pairwise comparison matrix with the factors in each stratum. In general, the scale used for pairwise comparison is from 1 to 9.

Step 3: To evaluate the relative weights of decision-making factors, solve the eigenvalues of the matrix constructed in Step 2. Estimate the relative weights of decision-making factors and verify CR (Consistency Ratio) to measure the reliability of the respondent' professionality. In general, a relative weight is obtained by calculating importance using Saaty (1980)s eigenvector. Only when CR is less than 0.1, we consider that reliability is secured [21].

Step 4: Aggregate the relative weights of the decision-making factors to prioritize the factors in each level. Then the priority of the items to evaluate is decided. It provides the basic information for resource allocation and alternative choice.

3.2 The Function Classification of EW System and Research Model

• The Function Classification of EW system

To calculate the relative weights on the areas of EW where drones have high usability, it is important to build a logic tree. When the logic tree is composed of systematic, simple, and essential contents, it enables the expert groups that participate in the survey to compare the weights of EW system logically and objectively in deciding them. Considering that the purpose of this study is to calculate the relative weights of the evaluation target, JCS Classification Standard of Chemical, Biological, and Radiological (CBR) EW Function was used in Table 4 to compose the logic tree.

• Sub-Classification of EW Functions [9]

(1) Table 5 shows the sub-classification of EW Functions.
(2) Table 6 shows the sub-classification of EA Functions.
(3) Table 7 shows the sub-classification of EP Functions.

• Research Model

Of the factors derived from the analysis in this paper, ES, EA, and EP were set as the factors in the high level. In the low level, they are divided as follows: ES into detecting, monitoring, locating, and identifying; EA into electronic jamming, electronic deception, anti-radiation guidance missile, and directed energy; and EP into anti-ES and anti-EA. AHP model was also constructed as seen in Table 8 to draw out the relative importance (priority) of the areas with a high usability of drones.

3.3 Expert Group and Questionnaire Collection

To calculate the weight of each factor means to quantify the priority of it. In this study, questionnaires were distributed and collected to the expert group to know the priority (weight) by factor based on the experts' subjective judgement.

• Questionnaire Design

The questionnaire was designed to ask the respondents to give a priority to the factors through the pairwise comparison of the factors in each stratum. Here, the factors by stratum are presented in following pages (the results of calculated weights). For the scale of measurement, a 9-point scale suggested by Saaty was used, as seen in Table 9. [1: Same, 3: Somewhat Important, 5: Important, 7: Very Important, 9]: Absolutely Important (2, 4, 6, and 8): the degree of importance that falls between [16].

(1) Weight on the sector of EW where a drone is highly usable.
      Please read the instruction in the table below, compare a set of two criteria, and mark a check in the box where you think of 'more important'.

Table 4 Classification standard of EW function by JCS [9]

Classification	Name of function	Description
EW (EW: Electronic Warfare)	Electronic Warfare Support (ES) (ES: Electronic Support)	It is all sorts of military activity to detect, analyze, and identify the electromagnetic energy that an enemy radiates. That is, ES is to detect, monitor, locate, analyze, and identify the electromagnetic energy emitting from the enemy's communication and non-communication; support a commander with information necessary for his or her decision; provide information needed for EW operations (EA, EP) and tactical activities (strike, counterintelligence (CI), combat damage assessment); and examine our army's EA and EP effectiveness. Tactical locating and tactical threat warning are included in ES
	Electronic Attack (EA) (EA: Electronic Attack)	EA is an activity to apply electromagnetic or directed energy to an enemy's troops, equipment, and facilities in order to disturb and neutralize its C4I system or degrade and destroy combat capability. In general, EA is operated with ES equipment in tactical level
	Electronic Protection (EP) (EP: Electronic Protection)	EP is all sorts of military activities to protect our army's human resources, facilities, and equipment whose degradation and destruction can lead to the neutralization of our army's combat capability by an enemy or our own EW operation. It protects our army's use of electromagnetic waves from the enemy's EW activities and electromagnetic interference and secures the effective operation of our electronic weapons system and C4I system. EP is divided into Anti-EA and Anti-ES. They are all sorts of military activities to inhibit the enemy's ES so that the effect of its activity to collect our army's intelligence can be minimized

Table 5 Electronic Warfare Support (ES)

Classification	Sub-classification	Description
Electronic Warfare Support (ES) (ES: Electronic Support)	Detection/Monitoring	It is the act to find out an enemy's communication and non-communication frequencies, listen to and record the contents of the detected frequencies and the resources of wave
	Locating	It is the act of locating the direction the enemy's electronic emitter arrives
	Identifying	It is the act to analyze the monitored and located contents, detect a tactical position, and give a warning

- Expert Group

To calculate the priority by factor for the usability of EW system mounted on a drone, 12 experts were selected who directly engage themselves in the field of EW for more than 10 years and qualified to give an objective evaluation.

4 Analysis Results

In this paper, Super Decision 3.0 and Excel were used to calculate and analyze consistency and weights in AHP model. In addition, when determining which areas have weights in case of using drones mounting EW system, CR was used: when it was below 0.3, it was determined there was consistency, but the questionnaire where CR exceeds 0.3 in analyzing weights was excluded.

Of the expert group of 12 with more than 10 years of service period in EW field, 8 experts turned out to give overall CR lower than 0.3. As seen in Table 10, 8 experts responded to pairwise comparison items and give weights.

Thomas L. Saaty said that arithmetic mean and geometric mean could be used to aggregate the importance that the experts thought of [15]. When it was determined that the experts involving in a decision-making process possess a high level of expertise, weighted arithmetic mean could be used after considering their weights. Geometric means are calculated by geometrically averaging out the values of the equal components (entries) of a pairwise comparison matrix. With them, a new general matrix can be constructed and weight can be obtained from this matrix. It has been proved that geometric mean is better in terms of the accuracy and consistency of a model in case the opinions of various experts are reflected at once [17].

The results of the first analysis demonstrated that ES was the more important than EA and EP among the factors in the upper level. Additionally, in the lower

Table 6 Electronic Attack (EA)

Classification	Sub-classification	Description
Electronic Attack (EA) (EA: Electronic Attack)	Electronic Jamming (Electronic Jamming)	It is the activity to apply unwanted electronic signals to electronic equipment by sending or sending back electromagnetic waves on purpose to degrade or disturb a receiver's reception of electromagnetic waves. It can be divided by implementation into electronic jamming, which actively radiates jamming electromagnetic waves, and mechanical jamming, which disturbs the enemy mechanically. The success of electronic jamming depends on the output of jamming equipment, the distance from a target, geographical features, and the type of an antenna
	Electronic Deception (Electronic Deception)	• Imitative Electronic Deception (IED): it is to imitate an enemy and get directly in its communication network to distribute false information • Manipulative Electronic Deception (MED): it is to allure the enemy to believe that false information is true so it can make important tactical mistakes or waste the assets of electronic warfare radiation • Simulative Electronic Deception (SED): it is to radiate electromagnetic signals to camouflage our army's electronic communication system
	Anti-Radiation Guidance Missile (Anti-Radiation Guidance Missile)	It is the missile that detects radio signals sent from a radar, track and destroy the radar
	Direction Energy (Direction Energy)	It is the weapons system to concentrate electromagnetic energy on a target and destroy, neutralize, or degrade. These types of weapons include laser, radio frequency (RF), and particle beam. They are expected to be potent EW tools in future battlefield

Table 7 EP

Classification	Sub-classification	Description
Electronic Protection (EP) (EP: Electronic Protection)	Anti-ES (Anti-ES)	It is all sorts of military activities to inhibit the enemy's ES so that the effect of its activity to collect our army's intelligence can be minimized • EMCON: it is the act of controlling and adjusting unnecessary electromagnetic radiation • Avoidance: it is the act of avoiding an enemy's electric wave detection by escaping from the range of an enemy's monitoring or reducing output radiation • It is to avoid the enemy's EW through spatial, temporal, and/or technical method such as controlling frequency in use, using jargons or codes • Electronic Camouflage: it is the act of radiating false electronic signals or transforming them so as to camouflage a target. (radio of crypto system and/or hopping system)
	Anti-EA (Anti-EA)	EA is an activity to apply electromagnetic or directed energy to an enemy's troops, equipment, and facilities in order to disturb and neutralize its C4I system or degrade and destroy combat capability. In general, EA is operated with ES equipment in tactical level • Electronic Jamming: It is the activity to apply unwanted electronic signals to electronic equipment by sending or sending back electromagnetic waves on purpose to degrade or disturb a receiver's reception of electromagnetic waves. It can be divided by implementation into electronic jamming, which actively radiates jamming electromagnetic waves and mechanical jamming, which disturbs the enemy mechanically. The success of electronic jamming depends on the output of jamming equipment, the distance from a target, geographical features, and the type of an antenna • Electronic Deception – Imitative Electronic Deception (IED): it is to imitate an enemy and get directly into its communication network to distribute false information – Manipulative Electronic Deception (MED): it is to allure the enemy to believe that false information is true so it can make important tactical mistakes or waste the assets of electronic warfare radiation – Simulative Electronic Deception (SED): it is to radiate electromagnetic signals to camouflage our army's electronic communication system • Anti-Radiation Guidance Missile: It is the missile that detects radio signals sent from a radar, track and destroy the radar

Table 8 Electronic warfare classification [9]

The hierarchical structure of drone usability in EW by its' priority

ES (Electronic Warfare Support)	EA (Electronic Attack)	EP (Electronic Protection)
• Detection/monitoring • Locating • Identifying	• Electronic jamming • Electronic deception • Anti-radiation guidance missile • Direction energy	• Anti-ES • Anti-EA

level of ES, identifying (0.254102), detecting/monitoring (0.249869), and locating (0.136333) turned out to be important factors in the order. Table 11 shows the results of the first analysis.

Since the experts in the field of EW keep performing their specialties (job functions) without interruption or changing. Therefore, it seemed necessary to know the difference in weight they estimated depending on their service year in the job. With 20 years as the reference service year, 6 experts were chosen with service years longer than 20 years and 6 experts shorter than it and conducted additional analysis. The results are in Table 12. According to the results of the second analysis, both experts groups (with the service period longer and shorter than 20 years) thought ES was important for drones, but 'below 20 years' group determined EP was more important for drones than EA and 'above 20 years' group looked at the question the other way around. In the difference of sub-factors, ES didn't show any change in the order of priority, but 'above 20 years' group thought of detection and monitoring more important than 'below 20 years' group. They showed a considerable difference in weight for Anti-AS under AP. 'Below 20 years' group thought of Anti-AP (0.1546) as the third important factor while 'Above 20 years' group thought of it as important as 0.0560.

5 Conclusion

In the 4th Industrial Revolution, drone industry has come into a spotlight as a national strategic industry all around the world. It is also the case for military sectors where drones are expected to play important roles. Studies for drones have already been carried out in various areas and they are expected to keep expanding their presence to more fields. However, not much has been studied in EW sectors as an intelligence asset, so we do not have sufficient technology and data. This paper has significance in that it is a study of drones as a way to overcome the weak reliability of EW system in locating, identifying, and etc. It seems the priority by factor for EW system when the system is mounted on a drone has an excellent acceptance of 'fused' technology: ES was the most important factor for a drone when mounting EW system, which was

Table 9 The example of the questionnaire

Criterion A	Important								↔	Important								Criterion B
	9	8	7	6	5	4	3	2	1	2	3	4	5	6	7	8	9	
ES	9	8	☑	6	5	4	3	2	1	2	3	4	5	6	7	8	9	EA
ES	9	☑	7	6	5	4	3	2	1	2	3	4	5	6	7	8	9	EP
EA	9	8	7	6	☑	4	3	2	1	2	3	4	5	6	7	8	9	EP

Table 10 8 items with overall CR below 0.3

Classification	1	2	3	4	5	6	7	8
Occupation	Soldier	Soldier	Soldier	Soldier	Soldier	Soldier	Soldier	Soldier
Gender	Male	Male	Male	Male	Male	Male	Male	Male
Age	47	43	39	49	49	44	41	47
Academic achievement	High school graduate	College graduate	High school graduate	University graduate	University graduate	High school graduate	College graduate	College graduate
Service years	25	21	17	28	28	24	20	26
Rank	Sergeant major	Sergeant major	First sergeant	Sergeant major	Sergeant major	First sergeant	First sergeant	Warrant officer
One's specialty	EW	EW	EW	EW	EW	EW	EW	EW
Overall consistency	0.00	0.01	0.25	0.30	0.24	0.02	0.13	0.20

Table 11 The results of the first analysis

Upper level	Weight	Lower level	Weight	Order
Electronic Warfare Support (ES)	0.6403	Detection/monitoring	0.249869	2
		Locating	0.136333	3
		Identifying	0.254102	1
Electronic Attack (EA)	0.22971	Electronic jamming	0.048565	7
		Electronic deception	0.039951	9
		Anti-radiation guidance missile	0.051134	6
		Direction energy	0.09006	4
Electronic Protection (EP)	0.12999	Anti-ES	0.083608	5
		Anti-EA	0.046378	8

followed by EA and EP. When a drone is mounted with EW system for the purpose of ES, identifying (0.254102) was most important in ES sub-factors, which were followed by detection/monitoring (0.249869), and locating (0.136333).

The additional analysis also showed both expert groups (above and below 20 years of service period in EW sector) gave much higher weight to ES than to EA and EP. It can imply they think when drones are used for EW system, it is easy to detect early and identify an enemy's activities and thus take a protective measure. In addition, the high weight of 'identifying', which is the sub-factor of ES represents the experts' expectation that technology for multiband detection and analysis will be improved along with the 4th Industrial Revolution and the development of IT technology. Comparable, the weight of Anti-ES, which is the sub-factor of EP, turned out to be higher than Anti-EA. It explains that the experts think EP will be more important in inducing an enemy's ES activities easily and detecting its threat early by controlling or adjusting radar and communication equipment in current use besides EW system.

It is admitted that this paper has some limitations: the weights for the factors cannot be absolute because the items for evaluation and the samples are limited. In other words, when respondents change, it can lead to different weighting. Therefore, there is a need for following studies to expand the items of evaluation and the samples. Nonetheless, this paper provides academic implications which factors of EW system should be considered first when a middle- and long-term policy is established in relation with drones and EW system, which are intelligence assets in the military sectors.

Last, as mentioned early, many countries place to top priority on drone industry, designating it as a strategic industry. In this respect, this study used AHP analysis to identify the factors to be considered when EW system is mounted on drones and prioritized the factors. As the military sectors are becoming more active to do R&D on and acquisition business with drones, it is expected that more extensive studies will be conducted to acquire quantitative and objective factors.

Table 12 The results of the additional analysis

Classification	Below 20 years	Above 20 years	Variable	Below 20 years		Above 20 years	
				Weight	Order	Weight	Order
ES	0.6718	0.7898	Detection/ monitoring	0.1162	4	0.2041	2
			Monitoring	0.1765	2	0.1802	3
			Identifying	0.3790	1	0.4054	1
EA	0.1397	0.1419	Electronic jamming	0.0273	8	0.0244	6
			Electronic deception	0.0508	5	0.0134	8
			Anti-radiation guidance missile	0.0342	6	0.0211	7
			Directed energy	0.0273	8	0.0829	4
EP	0.1883	0.0682	Anti-ES	0.1546	3	0.0560	5
			Anti-EA	0.0337	7	0.0122	9

References

1. Electronic Warfare, Joint Publication 3-13.1, I5-I18 (2012, 2007)
2. Bang, G.S., Lim, J.S.: Current status and development trend of maritime electronic warfare. Def. Technol. **223**, 56–63 (1997)
3. https://www.lockheedmartin.com (2018)
4. https://ko.wikipedia.org (2018)
5. https://ledger.wikispaces.com (2018)
6. http://www.victek.co.kr (2018)
7. Lee, K.L., Jang, H.J.: Information and electronic warfare sector. Sci. Technol. Vis. Def. Res. Dev. (3) (2007)
8. Yoo, T.S., Lee, G.I., Yoo, S.C.: Direction of development of electronic warfare system in preparation for the future. Def. Technol. **334**, 32–47 (2006)
9. http://medcmd.mil.kr (2018)
10. Kim, S.I., Jang, W.J., Ju, C.W., Lee, K.H., Kim, H.C.: Technical trend of defense electronics electrical IC for defense. Anal. Electron. Commun. Trends **24**(6) (2009)
11. http://www.kdrone.org (2018)
12. Kim, S.B., Kim, S.J.: Trends and implications of unmanned aircraft for major military units. Weekly Def. Bull. Korea Inst. Def. Anal. (KIDA) **1501**(14-6) (2014)
13. Han, D.I., Lee, D.W.: Unmanned Aircraft, Robots and Humans, pp. 10–18 (2012)
14. Unmanned System Integration Roadmap FY2013–2038, Defense Agency for Technology and Quality (DTaQ). Military Trend Series 14-02, 20–22 (2014)
15. http://www.ajunews.com (2018)
16. Woo, C.S., Kim, K.Y., Kang, S.B.: A comparative study of bankruptcy prediction model using LOGIT and AHP. Korea Financ. Manag. Assoc. **2**(2), 229–252 (1997)
17. Bin, J.Y.: (An)Analytic Network Process Model for Analyzing the Risk Factor of Water Supply Network. Korea University, Master Paper (2005)
18. Kim, K.Y.: The use of AHP for project management. PROMAT **7**(2) (1997)
19. Park, H., Ko, G.G., Song, J.Y., Shin, K.S.: A plan of multi-criteria analysis for preliminary feasibility study. Seoul. Public and Private Infrastructure Investment Management Center, The Korea Development Institute (2000)
20. Saaty, T,L.: The Analytic Hierarchy Process. McGraw Hill, New York (1980)
21. Zahedi, F.: The analytic hierarchy process—a survey of the method and its applications. Interfaces **16**(4), 96–108 (1986)

A Mechanism for Online and Dynamic Forecasting of Monthly Electric Load Consumption Using Parallel Adaptive Multilayer Perceptron (PAMLP)

J. T. Lalis, B. D. Gerardo and Yungcheol Byun

Abstract The study is based on time series modeling with special application to electric load consumptions modeling in a distributed environment. Existing theories and applications about artificial neural networks, backpropagation learning method, Nguyen-Widrow weight initialization technique and autocorrelation analysis were explored and applied. An adaptive stopping criterion algorithm was also integrated to BP to enable the ANN to converge on the global minimum and stop the training process without human intervention. These algorithms were combined together in designing the parallel adaptive multi-layer perceptron (PAMLP). In the simulation, the electric load consumptions of the seven (7) power utilities from Alaska in 1990–2013 were obtained from the official website of U.S. Energy Information Administration. The data set was divided into three overlapping parts: training, testing and validation sets, based on the principles of sliding-window training and walk-forward testing methods. The PAMLPs were trained and tested using the sliding-window method with 15-year window size and walk-forward testing method, respectively. The accuracy of each forecasting model produced by PAMLP was then measured using the respective out-of-sample validation sets using RMSD, CV (RMSD), and SMAPE ($0\% \leq SMAPE \leq 100\%$). In the monthly basis time series forecasting, the average CV (RMSD) at 7.79% and SMAPE at 3.12% for all utilities show the effectiveness of the PAMLP system across different time horizons and origin.

Keywords Dynamic forecasting · Monthly electric load consumption
Multi-layer perceptron

J. T. Lalis
College of Computer Studies, La Salle University, Ozamiz City, Philippines
e-mail: jeremias.lalis@lsu.edu.ph

B. D. Gerardo
Institute of ICT, West Visayas State University, Lapaz Iloilo City, Philippines
e-mail: bgerardo@wvsu.edu.ph

Y. Byun (✉)
Department of Computer Engineering, Jeju National University, Jeju, Korea
e-mail: ycb@jejunu.ac.kr

© Springer Nature Switzerland AG 2019
R. Lee (ed.), *Computer and Information Science*, Studies in Computational
Intelligence 791, https://doi.org/10.1007/978-3-319-98693-7_12

1 Introduction

Rapid growth of population and economy comes with rapid increase in the demand for electricity. However, this shows a positive impact as it signifies improvement in the standards of living of the people and growth of industries. Thus, planning for the production of electric power is very important to ensure that the supply will consistently meet the demand for electricity. Electric power production planning is done based on forecasts of future power load. There are several ways to do the forecasting and Box-Jenkins method is one of the most widely used time series forecasting method [1]. However, some studies [2, 3] show that the use of artificial neural networks (ANN) outperformed the other methods since it produces more accurate forecasting models. Furthermore, the capability of ANN to model non-linear relations even in the absence of human experts' supervision makes it more advantageous compared to the other methods [4]. To date, ANNs have been widely used to forecast monthly electric load consumption through trend extrapolation.

However, the complex nature of monthly electric energy consumption curve makes forecasting through trend extrapolation difficult to do. In general, consumption curve consists of two subtrends, the long-term increasing trend and the periodic waves. Long-term increasing trend occurs when there is a significant improvement in the nation's socio-economic aspects. On the other hand, periodic waves occur as the regularly rotating seasons drive people to change their work styles and living habits. The differences and presence of these two subtrends greatly affect the ability of ANNs to generalize. As a result, ANNs produce forecasting models with acceptable accuracy but with poor forecasting precision [5]. In order to overcome this difficulty, [6] proposed a method to improve the precision of the ANN model through feature extraction. And in the study of [5], the trend extracted through their proposed method was used to train the Radial basis function (RBF) neural networks to forecast the monthly consumption of China. The proposed methods exhibited good performance based on test results.

But there is another way to produce accurate and precise forecasting models even with the presence of these two subtrends. In developing ANN-based forecasting model, other influencing factors such as: temperature and humidity [7] and weather [8] records are used along with the monthly electric load consumptions. Yang et al. [9] also combined the economic and demographic variables, and historical load. Unfortunately, acquiring the records of these influencing factors on real-time manner is difficult to do, especially in a monthly basis, due to its incompleteness or unavailability. Real-time and on-line forecasting of monthly electric energy consumption plays a significant role in having an efficient and economic operation in the electric power utility. Forecasts will help the electric utility to make crucial and important decisions in generating/purchasing electric power, load switching and even in developing infrastructure. However, most of the existing studies focused on building static models trained in an off-line manner [9]. This is probably due to the difficulty associated in the determination of the appropriate architecture for the ANNs. Generally, the number of input nodes, hidden neurons, and even the stopping criterion are obtained through "trial-and-error" method. This process is very important since an ANN-based prediction model can produce better outcome than the other more complex models if it is just appropriately tuned [10].

Therefore, this study proposes an adaptive multilayer perceptron (AMLP) to obtain accurate and precise forecasting models even with the presence of subtrends and absence of influencing factors. Furthermore, this method is designed to support on-line and real-time training, thus, increasing its applicability in electric power production planning.

2   Review of Related Literature

2.1   Multilayer Perceptron and Backpropagation Learning

Artificial neural network (ANN), specifically the multilayer perceptron (MLP), is the simplest but widely accepted type of neural network in modeling complex real-world problems. MLP's architecture is composed of three interconnected layers known as: input layer, hidden layer/s, and output layer. Independent variables, termed as patterns, are assigned to the input layer nodes. On the other side of the MLP, the output on each neuron in the output layer represents the dependent variable. The hidden layer neurons, connecting the input nodes and output neurons through weighted links, capture the linear or even non-linear relationship between the independent and dependent variables. MLP's architecture is so versatile that it is even used in the field of forecasting.

In order to build a forecasting model, MLP should be trained first using significant amount of historical data. In general, ANNs can be trained either by supervised or unsupervised learning method. In this study, only one specific type of supervised learning method are being used, the backpropagation (BP) learning method. At the beginning of the training process, neurons in the hidden and output layers are initialized to small pseudo-random values.

BP algorithm then uses pairs of input and output patterns from the learning set and calculate the error between them. The resulting errors trigger the adjustments in the neuron weights. This process is done in order to minimize the error rate between each pair. The training will continue, except for weight initialization, until the network satisfies the selected stopping criterion.

2.2   Autocorrelation Analysis in Time Series

It has been proven that the number of nodes in the input layer has a significant impact in the forecasting performance of MLP [11]. Being able to identify the appropriate number of input nodes means better forecast outcome. However, traditionally, it is done manually through trial-and-error method. But the identified number of input nodes may not perform well in other time series due to the complexity of the consumption curve. In order to solve these problems, the researchers used the autocorrelation analysis to automate and effectively identify the appropriate number of input nodes based on the presented time series window. The correlation between xi and xi+k is computed based on the following equation:

$$r_k = \frac{\sum_{i=1}^{n-k} \frac{(x_i - \bar{x})(x_{i+k} - \bar{x})}{n-k}}{\sum_{i=1}^{n} \frac{(x_i - \bar{x})^2}{n}} \tag{1}$$

2.3 Nguyen-Widrow Weight Initialization

BP algorithm requires huge amount of processing power and longer time during the training phase. In order to significantly decrease the training time, initial weights of hidden layer neurons are calculated by integrating the Nguyen-Widrow randomization technique [12] in the standard BP. Through this algorithm, the trainability of the network has sufficiently increased, thus, resulting to a faster training process. Initial weights are calculated based on the following algorithm:

1. Assign the hidden and output neuron weights with small values using any pseudorandom number generator.
2. Use the number of input and hidden neurons, i and h respectively, to compute the Beta β using the equation:

$$\beta = 0.7h^{\frac{1}{i}} \tag{2}$$

3. Calculate the Euclidean norm n of all hidden neuron weights with:

$$n = \sqrt{\sum_{i=0}^{i<w_{max}} w_i^2} \tag{3}$$

4. Once the β and n are obtained, all weight values in the hidden layer are adjusted as follows:

$$w_{t+1} = \frac{\beta w_{t+1}}{n} \tag{4}$$

2.4 Adaptive Stopping Criterion

Choosing a stopping criterion to stop the BP learning process should be done carefully to ensure the model's optimal forecasting capability. Doing this in a trial-and-error method is not practical and feasible since the MLPs will be presented to different time series windows and consumption curves. In order to automatically and dynamically identify the stopping point of the training process, an adaptive stopping criterion [13] was integrated to the standard BP algorithm.

3 Distributed Parallel Adaptive Multi-layer Perceptron Model

One of the main characteristics and drawbacks of artificial neural networks is the time, which is associated with its computational complexity, it require during training phase. This is true especially on large volume of data set. The simultaneous learning of parallel AMLPs using the assigned training set will greatly reduce the time required in producing the predictive model. This approach is called training set parallelization.

Data-parallel architecture was adopted in this approach, wherein the entire AMLP is copied on the memory of each workstation. A portion of the entire training set is then assigned per workstation and trained in a full sequential manner. All the workstations must be trained simultaneously in order for the entire network to learn the whole feature space even though the data sets are divided into smaller parts and assigned to different workstations. To do this, the proposed method will be implemented on a single instruction, multiple data (SIMD)-based system.

In SIMD architecture type, all the machines execute the same program in a loosely synchronous fashion but on different data space parts. The parallel execution of program is controlled by the central machine using a single thread of instruction. Electric loads are produced by power plants and carried by high voltage transmission lines to different power substations. The substation then distributes the power to its residential, commercial and industrial consumers. The electric cooperative/utility on each substation will then record the monthly electricity consumption and collect the payments of their respective consumers.

In the proposed system, software with PAMLP will be installed on the computers of the distributed electric cooperatives or utilities. The monthly consumption of each electric cooperative will be used as input values to the system. The parallel workstations with PAMLPs will be trained in a loosely synchronous fashion based on the schedule determined by the central control. The central control will then save all the resulting predictive models from the parallel workstations on its database. The forecasted medium to long-term electricity requirements will be posted in an online website in a "per-substation" or "all-substations" form. With this method, different power industries and even the government can use the real-time and online electric load forecasting system as a tool in making decisions to ensure sustainable energy production and consumption.

3.1   Three-Tier Architecture

The 3-tier architecture aims to achieve system flexibility, robustness, scalability and resistance to potential changes in the environment. The architecture is composed of three independent layers: the presentation tier, the logic tier, and the data tier. The presentation tier provides the graphical user interface (GUI) to the end users. On the other hand, the logic tier provides application services to the end users, wherein, calculations and coordination between tier-1 and tier-3 are being done. Finally, data modeling, storage and management are conducted in the data tier.

The three-tier architecture aims to achieve system flexibility, robustness, scalability and resistance to potential changes in the environment. The architecture is composed of three independent layers: the presentation tier, the logic tier, and the data tier. The presentation tier provides the graphical user interface (GUI) to the end users. On the other hand, the logic tier provides application services to the end users, wherein, calculations and coordination between tier-1 and tier-3 are being done. Finally, data modeling, storage and management are conducted in the data tier.

In this study, the application, as a whole, is designed to strictly follow the rules of the 3-tier architecture. Wherein, the codes on each layer are contained separately to its specific locations, and thus, developed and maintained separately. Furthermore, each layer is totally unaware of the inner workings of the other layers and performing only its specific function. In this way, adding/modifying new/existing functionality on the specific layer will not necessarily have an effect on the other existing components.

3.1.1   Presentation Tier

Due to the application's architecture, large amount of users can access the system at the same time through the use of any available browsers, such as Mozilla Firefox, Internet Explorer and etc. The web page is designed and developed using PHP, JavaScript and Google Chart through AngularJS framework. The combination of this web and web-scripting technologies produced a dynamic web page with flexible content generation. Through this, the user can easily interact with the system by just simply entering the ID or name of the specific utility in the search box and view the forecasted monthly values by providing the necessary information such as start and end date.

Note that this tier has no direct communication with the data tier. As the end-user sends a request through the GUI, the request will be then forwarded to the logic tier's application server. This server will then send another request to the data tier's server. After processing the request, the response will be sent back to the logic tier in a JSON format. The logic tier will then process the response by parsing it and produce the graph using the Google Chart API. The graph is then displayed to the end-user's browser.

3.1.2 Logic Tier

In this study, the logic tier is divided into two parts. The first part, the application server, is responsible for bridging the gap between the GUI and the underlying database. This specific task is performed by the Apache server that is located in the logic tier. Note that the Apache server has no direct access to the database. It can only send requests to, and receive responses from, the data tier server (Oracle APEX Web Listener). The received response will be then processed and forwarded to the presentation tier by the logic tier's server.

The second part is responsible for calculating the future monthly electric load consumptions of each utility through PAMLP. The system is designed to be platform independent so that it can be easily distributed and installed to utilities' computers. Once it is installed, the user needs to provide the information about the utility and its monthly consumption records through the GUI. The PAMLP then automatically communicates with the data tier for training schedule. Once it received the schedule, it automatically starts the training process, which is running on the background, and produce forecast models based on the inputted past monthly consumptions. The forecasted values are then forwarded to the data tier for storage, usage and management. Generally, the PAMLP is an improved version of feedforward neural networks that are trained using a supervised learning method, the backpropagation learning method. Different adaptive algorithms were developed and combined together in order to have a dynamic, real-time and online way of forecasting the electric load consumptions. The algorithms are autocorrelation analysis, which determines the appropriate number of input nodes; Nguyen-Widrow weight initialization technique [12], which speeds-up the training process; and the adaptive stopping criterion [13], which enables the system to produce optimal forecasting models even without human intervention.

Multi-layer perceptron use a variety of learning techniques and backpropagation is the most common supervised learning methodology used on it. In this study, adaptive algorithms were integrated with MLP to form the adaptive multi-layer perceptron (AMLP) as proposed in the study of [14] and presented in Fig. 1. During the training, the AMLP was assigned for each month to forecast the monthly consumption of energy as shown in Fig. 2.

The AMLP works as follows: autocorrelation method was used to determine the number of nodes in the input layer based on the presented time series. The inputted time series values were then scaled to values ranging from 0 to 1. The calculated number of input nodes and scaled data set were used to create the training and validating sets. During the training, the Nguyen-Widrow randomization technique was used to initialize the weights of the neurons in the hidden layer. This generates a more trainable MLP. Once the MLP architecture is completely defined, the networks are then trained three times for each hidden layer neuron. The training continues until it converges to the global minima.

Once the training is completed, the model with the least weight is selected. Once the best model is identified, a set of patterns will be entered and scaled to forecast the next month's electric load consumption.

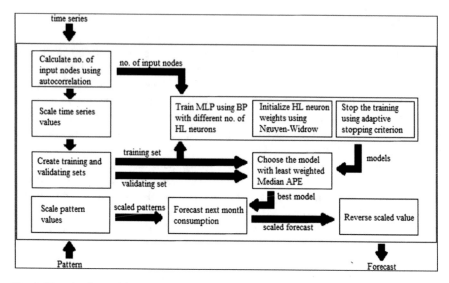

Fig. 1 The adaptive multi-layer perceptron (AMLP)

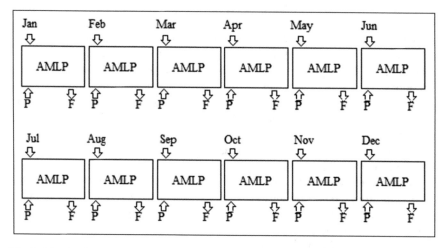

Fig. 2 Monthly forecasting scheme using AMLP

3.1.3 Data Tier

The forecasted electricity consumptions coming from the logic tier are being stored in the data tier. Aside from that, the information about the utility as well as the training schedules are also stored in this tier. End-users can then access this information in the presentation tier in a form of reports or graphs. To increase the flexibility and accessibility of the system, the Oracle Application Express (APEX) serves as the data tier and provides representational state transfer (REST) ful services to the logic tier. For every hypertext transfer protocol (HTTP) request sent by the logic tier, the data tier's web service application will return the response data in a JavaScript object notation (JSON) format. This data serve as an instruction to the distributed AMLP as when to start the training process in a parallel manner and forecast values.

4 Experiments and Results

4.1 Data Description

To test the robustness, accuracy, and precision of PAMLP, the electric power consumption (in megawatt-hour, MwH) data sets of the seven (7) different power utilities in Alaska from year 1990 to 2013 were used during the simulation. The data set was taken from the official website of U.S. Energy Information Administration [7]. For each utility, its corresponding data set was divided into three overlapping parts: training, validating and testing sets, with a 15-year time window. During the simulation, each utility where trained in a sliding-window manner.

For example in the window of 15 years, measurements of the electric consumption associated with the first 15 years were selected as the initial training data set. Once the initial training is done and the prediction for electric consumption for 16th year is generated, the electric consumption corresponding to the first year of the time series is discarded and the actual consumption in the 16th year is added in the window and retraining is carried out.

4.2 Simulation

The PAMLP was distributed and installed to ten (10) different personal computers in order to simulate the process of forecasting the annual and monthly electric consumptions of different power utilities. The first six (6) PCs represent the first three (3) utilities: Alaska Electric Light and Power Co. (01213 and 00213), Anchorage City of Alaska (01599 and 00599) and Chugach Electric Assn. Inc. (13522 and 03522), wherein, three of it are running under Linux Ubuntu platform and the other three (3) are replications of these utilities under Windows platform.

The rest of the PCs simulate the four (4) remaining utilities: Golden Valley Elec. Assn. Inc. (07353), Kodiak Electric Assn. Inc. (10433), Matanuska Electric Assn. (11824) and Homer Electric Assn. Inc. (19588), using the Windows platform. Figure 3 shows the sample forecasted monthly consumption of Chugach Electric Assn. Inc.

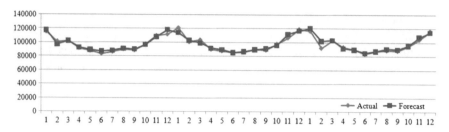

Fig. 3 Actual and forecasted monthly electric consumption of 13522/03522

4.3 Model Evaluation and Analysis

The prediction accuracy of the model on each simulation was measured using the root mean squared deviation (RMSD) and coefficient of variation of RMSD, CV(RMSD), denoted as,

$$RMSD = \sqrt{\frac{\sum_{t=1}^{n} \left[y_{pred}(t) - y_{data}(t)\right]^2}{n}} \qquad (5)$$

$$CV(RMSD) = \frac{RMSD}{|\bar{y}_{data}|} \times 100 \qquad (6)$$

where

n   the number of data,
ydata   the mean of the measured data,
ypred(t) the predicted energy at time t,
ydata(t) the measured data at time.

The CV (RMSD) is frequently used to describe how fit is the model in terms of outcome and squared residual values. Low CV means small residuals to the predicted value and thus suggestive of a good model fit. It also represents the ratio of the RMSD to the mean of actual data and can be used in comparing the degree of variation from one data series to another even if the means are drastically different from each other. CV (RMSD) was used in this study to determine the accuracy of PAMLP across different utilities and data set.

To further evaluate the performance of the algorithm, the symmetric mean absolute percentage error (SMAPE) was used to measure the performance of PAMLP. SMAPE is an accuracy measure based on percentage or relative errors which is defined as follows:

$$SMAPE = \frac{1}{2} \sum_{t=1}^{n} \frac{|F_t - A_t|}{(A_t + F_t)} (100) \qquad (7)$$

where A_t is the actual value and F_t is the forecast value.

Expression (7) was used to evaluate the PAMLP models as it does not depend on whether At is higher than Ft or vice versa. SMAPE also provides a well-defined range, $0 \le SMAPE \le 100\%$, to judge the size of relative errors between the actual and forecasted values.

4.4 Model Evaluation and Analysis

Based on Table 1, the PAMLP system performs best in monthly time series forecasting for utility 03522 based on CV (RMSD) at 3.31%. On the other hand, its worst performance is in utility 00213 with 14.71% variation. Note that utility 00213 is a replica of utility 01213. The average CV (RMSD) at 7.79% and SMAPE at 3.12% show the effectiveness of PAMLP by combining the adaptive methods of determining the number of nodes in the input and hidden layers, initialization of weights, and stopping criterion. These also prove that the proposed method can produce accurate forecasting models.

5 Conclusion

In this study, the researchers proposed a method to increase the accuracy and precision of the forecasting model generated through multi-layer perceptron (MLP). The monthly consumptions of the seven (7) power utilities in Alaska were used in training the parallel adaptive multilayer perceptron (PAMLP). Forecast values produced by the PAMLP models were also validated and tested for accuracy and precision. The integration of different adaptive algorithms such as serial correlation analysis, Nguyen-Widrow weight initialization and adaptive stopping criterion in the

Table 1 Monthly forecasting results summary using validation sets

Utility	RMSD	CV (RMSD) %	SMAPE %
01213	4,386.44	13.84	5.92
00213	4,660.36	14.71	5.95
01599	4,835.23	5.34	2.11
00599	4,241.18	4.68	1.90
13522	3,223.47	3.31	1.18
03522	3,964.66	4.07	1.40
07353	10,150.07	9.52	3.63
10433	1,059.61	8.88	3.46
11824	4,412.15	7.33	2.94
19558	2,509.46	6.25	2.67
Average	4,344.26	7.79	3.12

feedforward neural networks trained using backpropagation learning method has significantly decreased the training time and increased the forecasting capability of the produced models. Moreover, real-time and online building of dynamic forecasting models for monthly electric load consumptions has been realized. Based on empirical results, the PAMLP with eight (8) steps performs well in forecasting the electric load consumptions in monthly basis, in terms of accuracy and precision based on the resulting CV (RMSD) and SMAPE at 7.79% and 3.12%, respectively.

Acknowledgements This work (Grants No. C0515862) was supported by Business for Cooperative R&D between Industry, Academy, and Research Institute funded Korea Small and Medium Business Administration in 2017.

References

1. Ramarkrishna, R., Boiroju, N.K., Reddy, M.K.: Neural networks forecasting model for monthly electricity load in Andhra Pradesh. Int. J. Eng. Res. Appl. **2**(1), 1108–1115 (2012)
2. Kandananond, K.: Forecasting electricity demand in Thailand with an artificial neural network approach. Energies **4**, 1246–1257 (2011)
3. Edwards, R.E., et al.: Predicting future hourly residential electrical consumption: a machine learning case study. Energy Build. (2012). https://doi.org/10.1016/j.enbuild.2012.03.010
4. Haykin, S.: Neural Networks: A Comprehensive Foundation. Pearson Education Inc., Upper Saddle River, NJ, USA (2009)
5. Meng, M., Shang, W., Niu, D.: Monthly electric energy consumption forecasting using multi-window moving average and hybrid growth models. J. Appl. Math. (2014)
6. Meng, B.M., Niu, D., Sun, W.: Forecasting monthly electric energy consumption using feature extraction. Energies **4**(10), 1495–1507 (2011)
7. US Energy Information Administration. Monthly Energy Review. http://www.eia.gov/electric ity/data.cfm (2014). Accessed 26 May 2014
8. Kown, D., Kim, M., Hong, C., Cho, S.: Short term load forecasting based on BPL neural network with weather factors. Int. J. Multimedia Ubiquitous Eng. **9**(1), 415–424 (2014)
9. Yang, J., Rivard, H., Zmeureanu, R.: Building energy prediction with adaptive artificial neural networks. In: Proceedings of the 9th International IBPSA Conference, Montreal, Canada, pp. 1401–1408 (2005)
10. Arroyo, D., Skov, M., Huynh, Q.: Accurate electricity load forecasting with artificial neural networks. In: Proceedings of the 2005 International Conference on Computational Intelligence for Modelling, Control and Automation, and International Conference of Intelligent Agents, Web Technologies and Internet Commerce (2005)
11. Lin, F., Yu, X.H., Gregor, S., Irons, R.: Time series forecasting with neural networks. Complex. Int. **2** (1995)
12. Nguyen, D., Widrow, B.: Improving the learning speed of 2-layer neural networks by choosing initial values of the adaptive weight. In: Proceedings of the International Joint Conference on Neural Networks, San Diego, CA, USA, vol. 3, pp. 21–26 (1990)
13. Lalis, J.T., Gerardo, B.D., Byun, Y., Ha, Y.: Ubiquitous stopping criterion for backpropagation learning in multilayer perceptron neural networks. In: Proceedings of the 7th International Conference on Information Security and Assurance, Cebu City, Philippines, vol. 21, pp. 294–298 (2013)
14. Lalis, J.T., Maravillas, E.: Dynamic forecasting of electric load consumption using adaptive multilayer perceptron (AMLP). In: Proceedings of the International Conference on Humanoid, Nanotechnology, Information Technology, Communication and Control, Environment and Management, Palawan, Philippines (2014)

An Influence on Online Entrepreneurship Education Platform Utilization and Self-deterministic to College Students' Entrepreneurial Intention

Sung Taek Lee, Hoo Ki Lee, Hong Seok Ki and Gwang Yong Gim

Abstract 'Is it possible for entrepreneurship education in the university?' This question has been debated by many scholars over the years. The effect of entrepreneurship education through education is debatable from a form point of view [37]. Because the professor cannot give an analysis of the market in order to teach students his major and lead them to create new business from such major. On the other hand, there is also opinion that entrepreneurship can be trained and taught because it is an academic discipline, like any other field [12]. In order to solve such a controversy, it is necessary to establish entrepreneurship education at universities as a system that experts in the relevant field will conduct in a separate field from the major education. However, there is also a gap in the entrepreneurship education depending on the competence of the experts for entrepreneurship education. In order to solve such gaps in education, online education platforms such as MOOC are expected to be needed in entrepreneurship education.

Keywords Online entrepreneurship education platform · Entrepreneurial intention · Entrepreneurial Self-efficacy · Self-deterministic · Information system success model

S. T. Lee · H. K. Lee · H. S. Ki · G. Y. Gim (✉)
Department of IT Policy and Management, Soongsil University, Seoul, Republic of Korea
e-mail: gygim@ssu.ac.kr

S. T. Lee
e-mail: totona22@ssu.ac.kr

H. K. Lee
e-mail: hk0038@korea.kr

H. S. Ki
e-mail: redstone78@ssu.ac.kr

© Springer Nature Switzerland AG 2019
R. Lee (ed.), *Computer and Information Science*, Studies in Computational Intelligence 791, https://doi.org/10.1007/978-3-319-98693-7_13

1 Introduction

The university's entrepreneurship education has an important meaning in the broad meaning of pioneering a new field with a challenging spirit toward creating new value when designing its career through entrepreneurship cultivation. Recent trends in universities have revealed various policies to support young entrepreneurship, and they establish the system mentioned in the above as they start to recognize entrepreneurship education as an area of university education. However, the importance of ICT convergence education based on data science (Big Data, AI, Deep Learning, etc.) has emerged in the 4th Industrial Revolution era and development of technology based ideas, problem solving ability, demand for talented people with global capabilities is increasing. The education of the university is also changing in line with the 4th Industrial Revolution era. It is changing from the traditional majors to the interdisciplinary and convergent curriculum. In the current trend of change, the university's entrepreneurship education is changing from entrepreneurship-oriented education to characteristics of the 4th Industrial Revolution era and centered on cultivating start-up type talent through various experiences. However, there is also a gap in the entrepreneurship education depending on the competence of the experts for entrepreneurship education. And, major infrastructure related to start-up is concentrated in the metropolitan areas, causing the gap in regional education level and start-up support policy. In order to solve such gaps in education, online education platforms such as MOOC (Massive Open Online Course) are expected to be needed in the field of entrepreneurship education.

Therefore, this study will carry out research with the following research questions. First, this study is going to examine whether the use of online entrepreneurship education platform is useful to college student founders. Second, this study examines whether the quality characteristics of the contents provided by the platform can increase the self-efficacy, which is a personal characteristic factor of the user. Third, analyzes empirically the effects of platform characteristics on users' utility characteristics and users' efficacy on entrepreneurial intention. Finally, this study tries to analyze the difference of autonomy, competence, and relatedness of self-deterministic based on human psychological theory. The purpose of this study is to propose the design and operation plan of online entrepreneurship education platform to solve some of the problems that occur in university entrepreneurship education and to expand the effectiveness of university entrepreneurship education.

2 Entrepreneurship Education

2.1 Definition of Entrepreneurship Education

Entrepreneurship education can only be defined by focusing on the concept of business opportunity. Entrepreneurship education teaches students how to identify business opportunities, how to find start-up items, and how to analyze target customers [9]. Arranging above details, entrepreneurship education is simply beyond the scope of basic concepts and knowledge related to entrepreneurship [2]. In addition, entrepreneurship education is the development of skills and competencies necessary for start-up activities [9].

Based on these literature studies, this study is going to define entrepreneurship education as a process of transferring knowledge and skills related to start-up to students to utilize business opportunities. In other words, students must develop their knowledge and skills to solve the complex problems, risks and uncertainties inherent in corporate processes and to improve their attitudes toward entrepreneurship.

This entrepreneurship education applies to both those who want to be entrepreneurs and those who are not interested in becoming entrepreneurs. The group to be an entrepreneur is a typical target group of entrepreneurship education [26]. However, there is a need to participate in entrepreneurship education programs even if they are not interested in being an entrepreneur. This is because entrepreneurship education provides basic knowledge of entrepreneurship, improves entrepreneurial skills and innovation skills, and improves attitude toward entrepreneurship. Jeon (2012) emphasized the necessity of entrepreneurship education because the person who has participate in entrepreneurship education is active in establishing start-up plan and entrepreneurship education can change individual's psychological characteristics [18].

Taken together, above findings suggest that entrepreneurship education should focus on teaching entrepreneurial knowledge and skills and improving entrepreneurial attitudes and entrepreneurial intention. To do so, it is important to identify the types of knowledge and skills that entrepreneurship education programs must provide, and the factors that drive students to change their perceptions of the business.

2.2 Scope and Method of Entrepreneurship Education

Entrepreneurship education should emphasize theories and principles of entrepreneurship because it is useful to develop students' cognitive skills [14, 15]. However, researchers with contrary views have argued that a practically necessary, practice-based approach is more important [17]. Entrepreneurship teaching is important to both a theoretical as well as a practical aspect of entrepreneurship [1]. Founder's participation in systematic education programs is a key factor for successful start-up, and entrepreneurship education should consist of theory education and field practice. In particular, they emphasized the importance of field practice [23].

Entrepreneurship education should include general business knowledge, including market analysis and planning, pricing strategy, financial analysis, leadership, human resource management theory and other business theory and skills [4]. Some researchers have argued that entrepreneurship education, unlike general business education, should address issues related to business entry [16]. Students should be trained to make decisions in an uncertain environment, and that entrepreneurship education should focus on practical training on how to select and manage new businesses [33]. Collins et al. (2006) encourage cooperative learning among students in entrepreneurship education [8].

Traditional lecture-centered methods are appropriate for entrepreneurship education [40]. Entrepreneurship education requires practical learning based on experience rather than traditional lecture methods. In another study, we proposed a project method to conduct entrepreneurship education using business plan as a main tool [44].

In order to advance the entrepreneurship education, it is necessary to develop customized entrepreneurship education courses in stages, away from the one-off special lectures and theoretical education, and to compose the entrepreneurship and the commercialization process by finding ideas [19].

2.3   Effectiveness of Entrepreneurship Education

Scholars of entrepreneurship argued that evaluation of entrepreneurship education is important for developing effective entrepreneurship curriculum [4].

Many researchers focusing on new venture start-up have argued that there is a positive relationship between entrepreneurship education and start-up activity [16]. Entrepreneurship education positively influences on entrepreneurship and improves decision making [42]. Despite the positive effects of entrepreneurship education, only a small number of students will be involved in entrepreneurship education and start a business. The goal of entrepreneurship education is not that all participants create new business in a short period of time. It is important for students to have entrepreneurial attitude and entrepreneurial intention through entrepreneurship education.

Noel (2001) found that entrepreneurs who graduated from entrepreneurship education tend to be more willing to act with entrepreneurial intention and self-efficacy than business or other majors [29].

2.4   Online Entrepreneurship Education Platform

The online entrepreneurship education platform is to provide entrepreneurship education, mentoring, start-up support policies, and networks necessary for start-up activities in an online environment that is free from space and time constraints, so that universities can utilize them jointly.

Currently, the status of university start-up support is insufficient because there are not enough educators who have start-up experience in the university to provide practical start-up know-how needed to start-up challenge of college students. Major infrastructures related to start-up are concentrated in large cities, and there is a gap in regional education level and start-up culture. The online entrepreneurship education platform is designed to solve the gaps in the start-up support between universities and to improve the quality of start-up support projects such as supporting the experience necessary for practical start-up by increasing the utilization of the know-how and to provide equal opportunities for excellent entrepreneurship education.

The online entrepreneurship education platform will be built reflecting the characteristics of MOOC. MOOC provides a new learning environment that enables interactive learning such as Q&A, discussion, quiz, and task submission between instructor and learner, learner and learner, unlike the existing online learning system that the learner has only to listen to passively. The online entrepreneurship education platform will also be able to provide services such as business plan review and preliminary feasibility review from a teacher or expert group, as well as entrepreneurship education from the student's perspective. From the viewpoint of the instructor, student management service can be provided not only the lectures on entrepreneurship education but also by managing service the history of start-up activities of students.

3   Theoretical Background

3.1   Quality Characteristic of Information System Success Model

The quality of the information system is an important factor for effectively operating the information system. Pitt et al. [32] pointed out that the quality factors of the information system success model presented in the initial DeLone and McLean [10] study are limited to the system quality and information quality and In addition, service quality has been suggested as a success factor of information system.

Information quality is an important factor in determining the satisfaction of e-learning information system through studying success model of e-learning information system [20]. If the information provided in the information system is not accurate or reliable under an environment that there is enable overall exchange of communication between the professor and the student, data sharing, discussion, etc., the user will not be satisfied. In other words, accuracy and reliability are suggested as factors that evaluate information quality.

The results of a study on the effects of college students' satisfaction on entrepreneurship education are as follows [7]. First, the satisfaction of entrepreneurship education can be enhanced by strengthening empathy and differentiation among service quality factors of entrepreneurship education. Second, based on the results that service quality of entrepreneurship education affects entrepreneurial intention, the university needs activities to improve service quality in order to increase the entrepreneurial intention. Third, based on students' satisfaction with entrepreneurship education positively affects entrepreneurial intention, various efforts are needed to increase entrepreneurship education satisfaction of students.

Service quality of the online entrepreneurship education platform should be evaluated by focusing on the usefulness, supportability, and effectiveness of the start-up activity of the commercialization support programs necessary for the entrepreneurial activity in addition to the contents related to the entrepreneurship education provided by the online entrepreneurship education platform.

System quality is a technical characteristic of the information system itself that collects and processes accurate information and supports communication, and it means the technical quality that the user feels while using the system [10] as well as the system that deals with information (hardware, software, network, etc.) and the performance of the system itself, ease of accessing the system, flexibility and suitability of user requirements, system response/turnaround time, system.

3.2   User's Utility Characteristic

Perceived usefulness has a positive effect on user satisfaction through the analysis of the causal relationship between the quality characteristics of e-learning systems and learning outcomes. However, it was found that the quality characteristics of the system did not positively affect the perceived usefulness. Previous studies of the learning information system show that perceived usefulness has an inconsistent effect on usage intention and learning outcomes. The relationship with quality characteristics that affect perceived usefulness also has not been consistent. This is because it is not easy to judge the usefulness of the learner based on the experience of using the system and the clear definition of the voluntary or involuntary situation [43].

Yoo and Lee (2007) hypothesized that the quality characteristics of the learning information system would have a positive effect on user satisfaction, but it was deducted that the results of the study were not influenced. It can be said that user satisfaction was deducted with other result according to the voluntary and involuntary environment even for learners who use the learning information system similar to the perceived usefulness [43].

3.3   Entrepreneurial Self-efficacy

Start-up founders must achieve high levels of efficiency in start-up activities. The key factor to achieving this efficiency is entrepreneurial self-efficacy [41].

Self-efficacy, which is the internal factor of the start-up environment, means confidence. Some people are born with such self-efficacy, and others are acquired by growth environment and education. Entrepreneurship education is possible to contribute to establishment of acquired self-efficacy by education [30]. Research on self-efficacy in entrepreneurship education field has been proven to be an important predictor of entrepreneurial intention [27]. It is defined to be an important explanatory variable in determining entrepreneurial intention and it will lead to entrepreneurial behavior [5]. In particular, entrepreneurship education programs have been proven to be an important factor in increasing self-efficacy [31].

If students attended class related to entrepreneurship, or experienced start-up and had risky tendency influences on entrepreneurial intention by mediating entrepreneurial self-efficacy [45].

3.4 Self-deterministic

Deci and Ryan [11] refer that "self-deterministic is a qualification of human function that includes experience of choice. Self-deterministic is the ability of humans to decide and choose their own behavior without external interference or influence." In other words, it means the level which you feel you can adjust and control your own behavior.

Self-deterministic consists of three parts: autonomy, competence, and relatedness. In order to humans do self-determining behavior freely these three desires must be satisfied. The three needs that constitute the basic psychological desire are as follows: First, autonomy is a desire that people want to be a subject without any external control or pressure when they do something. Second, competence is a desire that people want to perform their assigned tasks successfully using their own abilities. It is a desire to exert their abilities and talents in interaction with the social environment and to act effectively [34, 35]. Third, relatedness is satisfied when one feels that one's own and others are forming an intimate relationship, that they feel that they belong to a certain society, and that they are being treated by others [22]. When the needs of autonomy and competence are satisfied, they want to use IT services continuously [21]. They argued that they use IT services for pleasure as they feel pleasure when need of relatedness with colleagues is satisfied.

3.5 Entrepreneurial Intention

There are many factors that influence the entrepreneurial intention such as personal propensity, career orientation, self-efficacy, start-up network, political support system, economic situation, and entrepreneurship education. This study researched about the characteristics of the online entrepreneurship education platform and the relationship between entrepreneurship education and entrepreneurial intention.

Self-efficacy, personal characteristics, autonomous propensity, and entrepreneurship education influenced on entrepreneurial intention positively [36]. This result will be a link to the study of moderating effects of autonomy that this study intends to analyze.

Students' self-efficacy had a significant impact on their entrepreneurial intention, and that entrepreneurship education which could increase self-efficacy could increase their entrepreneurial intention [27].

However, there are some studies that have resulted in contradictory analysis results. The relationship between entrepreneurship education and entrepreneurial intention is positive, irrelevant, or even negative [13].

4 Research Model and Hypothesis

4.1 Research Model

The characteristics of this research model are as follows: First, the existing researches focus on the introduction of information system success, while this study is going to examined what kind of structural relationship the personal characteristics (entrepreneurial self-efficacy) and disposition (self-determinism) forms in utilizing not only successful introduction of information systems but also entrepreneurship eduction platform. Second, this study is going to analyze the effect on user satisfaction and entrepreneurial intention through characteristics of users of information quality, service quality, and system quality of online entrepreneurship education platform. Third, this study analyzes the effect of user satisfaction on entrepreneurial intention by using online entrepreneurship education platform. Fourth, this study try to exclude the direct effects of quality characteristics of the information quality, service quality, and system quality of online entrepreneurship education platform on user satisfaction (Fig. 1).

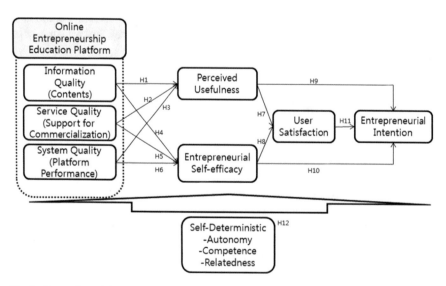

Fig. 1 Research model

4.2 Research Hypothesis

The purpose of this study is to examine the relevance weighing on entrepreneurial intention by establishing perceived usefulness and user satisfaction as utility factors, which is quality characteristics and personal characteristics of users, based on advanced studies in order to examine the effect of the online entrepreneurship education platform on the start-up activities of college students. In addition, the research hypothesis was established to examine the moderating effect of the self-deterministic constitutional factors (autonomy, competence, and relatedness), which is the tendency of the user in this relationship.

The hypotheses of H1, H2, and H3 are established assuming that the quality characteristics of the online entrepreneurship education platform have a significant effect on the perceived usefulness:

H1: The information quality of the online entrepreneurship education platform will have a positive (+) impact on perceived usefulness.

H2: The service quality of the online entrepreneurship education platform will have a positive (+) impact on perceived usefulness.

H3: The system quality of the online entrepreneurship education platform will have a positive (+) impact on perceived usefulness.

Since the online entrepreneurship education platform has the characteristics of the learning information system, this study examined the effects of the quality of the provided contents on personal characteristics. The quality of entrepreneurship education has a significant effect on personal characteristics such as self-efficacy and user satisfaction [24, 25]. The hypotheses of H4, H5, and H6 were established assuming that the quality characteristics of the online entrepreneurship education platform had a significant effect on the self-efficacy of entrepreneurial self-efficacy:

H4: Information quality of online entrepreneurship education platform will have a positive (+) effect on entrepreneurial self-efficacy.

H5: The service quality of online entrepreneurship education platform will have a positive (+) effect on entrepreneurial self-efficacy.

H6: The system quality of the online entrepreneurship education platform will have a positive (+) effect on the entrepreneurial self-efficacy.

Self-efficacy has a significant effect on learning satisfaction in mobile learning [28]. Self-efficacy has a significant effect on entrepreneurship education satisfaction [24]. Based on previous studies as above the hypotheses of H7 and H8 were established because it was judged that perceived usefulness and entrepreneurial self-efficacy had a significant effect on user satisfaction:

H7: Perceived usefulness of online entrepreneurship education platform will have a positive (+) effect on user satisfaction.

H8: Entrepreneurial self-efficacy of online entrepreneurship education platform will have a positive (+) effect on user satisfaction.

Seddon and Kiew [38] and Seddon [39] have shown that perceived usefulness and user satisfaction influence individual outcomes in information system success model studies. It was found that most of the variables did not influence on the relationship between variables, but only perceived usefulness and user satisfaction had a significant effect on learning outcomes, confirming that the effects of traits on individual performance were not found to be the same [43].

Boyd and Vozikis (1994) developed the theoretical framework of self-efficacy as an important prerequisite for entrepreneurship [5], and Chen et al. (1998) empirically proved the relationship between entrepreneurial self-efficacy and entrepreneurial intention [6]. H9, H10, and H11 were hypothesized as perceived usefulness, entrepreneurial self-efficacy, and user satisfaction on the basis of previous research results as above:

H9:   Perceived usefulness of the online entrepreneurship education platform will have a positive (+) impact on the entrepreneurial intention.

H10:   The entrepreneurial self-efficacy of the online entrepreneurship education platform will have a positive (+) impact on the entrepreneurial intention.

H11:   User satisfaction of online entrepreneurship education platform will have a positive (+) effect on the entrepreneurial intention.

The necessity and importance of entrepreneurship education is increasing as many scholars have shifted from the viewpoint that start-up founder can be developed from the perspective of being born. Although the results of research showing that start-up founder can be developed are shown, the relationship between quality characteristics, user utility characteristics, personal characteristics, and the entrepreneurial intention depends on the degree of self-deterministic factor (autonomy, competence, and relatedness). And this study intended to analyze whether there is a difference according to participation in entrepreneurship education.

H12   The relationship between quality characteristics, user utility characteristics, personal characteristics and entrepreneurial intention be different depending on the degree of self-determination and participation in entrepreneurship education.

5   Data Analysis

5.1   Characteristics of the Sample

A total of 517 questionnaires were conducted to remove the questionnaires with missing or inadequate answers, and the final 429 cases were selected as valid samples. Table 1 summarizes the sample of this study.

Table 1 Demographic characteristics of the sample

Category and items		Sample size	Ratio (%)
Gender	Male	268	62.5
	Female	161	37.5
Age	19–25	292	68.1
	26–30	77	17.9
	31–35	16	3.7
	36–40	14	3.3
	More than 41	30	7.0
Education level	1st grade	55	12.8
	2nd grade	67	15.6
	3rd grade	81	18.9
	4th garde and above	120	28.0
	University Graduation	62	14.5
	More than a master's course	44	10.3
Filed of major	Humanities	54	12.6
	Social Science	45	10.5
	Natural Science	46	10.7
	Economics & Business	83	19.3
	IT	60	14.0
	Engineering	86	20.0
	Arts & Physical Education	22	5.1
	Health & Bio	13	3.0
	Others	20	4.7
Times of participation in entrepreneurship education	None	212	49.4
	1 time	77	17.9
	2 times	46	10.7
	3 times	20	4.7
	4 times	9	2.1
	More than 5 times	65	15.2

Table 2 Conceptual validity evaluation result of measurement model

Item	IQ	SEQ	SYQ	PU	US	ESE	EI
CR	0.864	0.797	0.801	0.875	0.921	0.855	0.916
AVE	0.679	0.569	0.578	0.638	0.699	0.598	0.687

Table 3 Discrimination validity evaluation result of measurement model

	IQ	SEQ	SYQ	PU	US	ESE	EI
IQ	**0.824**						
SEQ	0.718	**0.754**					
SYQ	0.526	0.602	**0.760**				
PU	0.616	0.637	0.500	**0.799**			
US	0.361	0.446	0.324	0.722	**0.836**		
ESE	0.295	0.312	0.317	0.431	0.500	**0.773**	
EI	0.363	0.341	0.374	0.412	0.450	0.729	**0.829**

5.2 Analysis Result of Measurement Model

This study conducted hypothesis verification using Structural Equation Modeling (SEM).

Since composite reliability (CR) is 0.7 or more, and the average variance extracted (AVE) is 0.5 or more, it exceeds the threshold value of all the indexes and it can be judged that the conceptual validity of the measurement model is secured shown in Table 2.

Table 3 is composed as matrix of correlation between the latent variables in order to analyze the discriminant validity, and it can be evaluated to have secured discriminant validity as the largest correlation coefficient, 0.729 (correlation coefficient between entrepreneurial self-efficacy and entrepreneurial intention) is less than 0.754 (service quality), which is the smallest among the square root of average variance extracted (AVE).

5.3 Structural Model Verification Results

A result of structural model's fitness test, all of indices appeared above threshold criteria and therefore, the structural model' goodness of fit of the research model was verified in shown Table 4.

Table 4 Fit evaluation result of structural model

Fit index			Indicator value	Threshold criteria
Absolute fit index	Model overall fit	χ^2(p)	637.970	$p \leq 0.05$–0.10
			(P = 0.000)	
		χ^2/df	2.106	$1.0 \leq \chi^2/df \leq 3.0$
		RMSEA	0.051	$\leq$0.05–0.08
	Model explanation power	GFI	0.898	$\geq$0.8–0.9
		AGFI	0.873	$\geq$0.8–0.9
		PGFI	0.720	$\geq$0.5–0.6
Incremental fit index		NFI	0.917	$\geq$0.8–0.9
		TLI	0.947	$\geq$0.8–0.9
		CFI	0.954	$\geq$0.8–0.9
Simplicity Suitability Index		PNFI	0.792	$\geq$0.6
		PCFI	0.824	$\geq$0.5–0.6

Table 5 Hypothesis test result

Research hypothesis				Path coeffi-cient	Standard error	C.R. (t)	P	Test result
H1	IQ	→	PU	0.304	0.081	3.757	***	**Supported**
H2	SEQ	→	PU	0.427	0.113	3.784	***	**Supported**
H3	SYQ	→	PU	0.141	0.063	2.245	0.025*	**Supported**
H4	IQ	→	ESE	0.106	0.080	1.329	0.184	Rejected
H5	SEQ	→	ESE	0.136	0.099	1.375	0.169	Rejected
H6	SYQ	→	ESE	0.178	0.066	2.671	0.008**	**Supported**
H7	PU	→	US	0.582	0.046	12.757	***	**Supported**
H8	ESE	→	US	0.276	0.049	5.595	***	**Supported**
H9	PU	→	EI	0.119	0.061	1.952	0.051	**Supported**
H10	ESE	→	EI	0.842	0.079	10.621	***	**Supported**
H11	US	→	EI	0.061	0.071	0.864	0.388	Rejected

*$p < 0.05$, **$p < 0.01$, ***$p < 0.001$

5.4 Hypothesis Testing

In this study, we confirmed the structural equation modeling using AMOS to test hypotheses, verify relationships of key variables are shown in Table 5.

Table 6 Cross-validation analysis of autonomy difference groups

Division	Unconstrained model	Measurement weights model
χ^2	1026.895	1065.991
X^2 difference	39.097	
P-value of χ^2 difference	0.006	

5.5 Moderating Effects

This study conducted Multi Group Structural Equation Model Analysis (MSEM) to verify what mediating effect weighs on the relationship between the quality character-istics (information quality, service quality, system quality) and user utility character-istics (perceived usefulness, user satisfaction, entrepreneurial self-efficacy) depend-ing on high and low of self-deterministic factors. In addition, it was verified that what kind of control effect on the overall structural relationship is weighed by participation in entrepreneurship education.

The autonomy, competence, and relatedness were established as the factor to mea-sure the degree of self-determinants, and the average of respondents was calculated for each item in each sample, and the two comparative groups were divided in com-parison with the average of whole 429 respondents, and verified the cross validity through Multi-Sample Confirmatory Factor Analysis (MCFA) to determine whether the same analysis results could be obtained from each group extracted from the same population. This study verified whether it has overall mediating effect according to self-determinants degree in comparison with unconstrained model, with execution of multi-sample comparative analysis by structural weights model.

5.5.1 Moderating Effects by Autonomy

In order to confirm the moderating effects of autonomy, the cross validity of verifying the metric equivalence of competence between the lower (196) and higher (233) groups was verified. As a result of the cross validity analysis is meaningless to verify the moderating effect by autonomy by performing structural model analysis (Table 6).

5.5.2 Moderating Effects by Competence

In order to confirm the moderating effects of competence, the cross validity of verify-ing the metric equivalence of competence between the lower (211) and higher (218) groups was verified. A result of the cross validity analysis can be explained that the fitness of the model is not deteriorated and the cross validity between competence comparison groups is secured (Table 7).

Table 7 Cross-validity analysis of competence difference groups

Division	Unconstrained model	Measurement weights model
χ^2	1026.7	1057.77
X^2 difference	31.07	
P-value of χ^2 difference	0.054	

Table 8 Multiple comparison analysis between competence difference groups

Division	Unconstrained model	Structural weights model
χ^2	1066.153	1115.131
X^2 difference	48.978	
P-value of χ^2 difference	0.021	

It can be explained that measurement identity between competence comparison groups is secured, and the measurement weighted between competence comparison groups can be confirmed, having the result as in Table 8 with multi comparative analysis by competence group. It can be viewed that the χ^2 unconstrained and constrained model between two groups is 48.978 (1115.131 − 1066.153), p = 0.021 of χ^2 difference at 5% level which is statistically significant. Therefore, competence can be regarded as a controlling variable.

As this study confirmed that the moderating effect of competence was significant between the two groups, this study divided the sample into two comparative groups based on the high and low levels of competence, and analyzed the differences among the groups.

First, hypothesis that system quality affects perceived usefulness and entrepreneurial self-efficacy is rejected for low competence group, but high competence group is supported. This implies that the system quality of the online entrepreneurship education platform is considered to increase the usefulness of the start-up activity rather than the low group. In addition, it can be explained that the group with high competence has a high sense of efficacy on the online entrepreneurship education platform. Second, in a hypothesis that perceived usefulness affects entrepreneurial intention low-competence group was supported, but high-competence group is rejected. This can be interpreted that, in case of a high-competence group, those who feel that they have the competence to solve themselves, the usefulness upon online entrepreneurship education platform use can raise efficiency of the start-up activities, but, personal competence is affected more on entrepreneurial intention (Table 9).

Table 9 Moderating effect analysis result between competence groups

Hypothesis (path)			Low competence (211)			High competence (218)		
			Path coefficient (t)	p	Test result	Path coefficient (t)	p	Test result
IQ	→	PU	0.264 (2.393)	0.017*	**Supported**	0.334 (2.790)	0.005**	**Supported**
SEQ	→	PU	0.406 (2.903)	0.004**	**Supported**	0.351 (2.068)	0.039*	**Supported**
SYQ	→	PU	0.063 (0.769)	0.442	Rejected	0.214 (2.326)	0.020	**Supported**
IQ	→	ESE	0.000 (−0.004)	0.996	Rejected	0.047 (0.577)	0.564	Rejected
SEQ	→	ESE	0.049 (0.418)	0.676	Rejected	0.113 (1.074)	0.283	Rejected
SYQ	→	ESE	0.077 (0.924)	0.356	Rejected	0.132 (1.765)	0.078	**Supported**
PU	→	US	0.469 (7.694)	***	**Supported**	0.685 (9.228)	***	**Supported**
ESE	→	US	0.321 (3.706)	***	**Supported**	0.275 (3.139)	0.002**	**Supported**
PU	→	EI	0.210 (2.566)	0.010*	**Supported**	-0.030 (-0.331)	0.741	Rejected
ESE	→	EI	0.827 (5.602)	***	**Supported**	0.786 (6.225)	***	**Supported**
US	→	EI	0.010 (0.096)	0.924	Rejected	0.146 (1.562)	0.118	Rejected

*$p<0.05$, **$p<0.01$, ***$p<0.001$

5.5.3 Moderating Effect by Relatedness

In order to confirm the moderating effect on the relationship, this study verified the cross validity verifying the metric equivalence of the relationship between the group with lower group (224) and the higher group (205) than total average. A result of the cross validity analysis can be explained that the fit of the model after constraint that measurement weight is the same to unconstrained model is not deteriorated, and that the cross validity between the comparative groups is secured (Table 10).

It can be explained that the measurement equality is secured between the comparative groups, and this study confirmed that the measurement weights among the comparative groups are the same with multi comparative analysis by group (Tables 11).

Since this study confirmed that the moderating effect by relatedness was significant between the two groups, the sample was divided into two comparative groups

Table 10 Cross-validity analysis of relatedness difference groups

Division	Unconstrained model	Measurement weights model
χ^2	972.022	994.806
X^2 difference	22.784	
P-value of χ^2 difference	0.3	

Table 11 Multiple comparison analysis between relatedness difference groups

Division	Unconstrained model	Structural weights model
χ^2	1018.834	1065.979
X^2 difference	47.145	
P-value of χ^2 difference	0.032	

Table 12 Moderating effect analysis result between relatedness groups

Hypothesis (path)			Low relatedness (224)			High relatedness (205)		
			Path coefficient (t)	p	Test result	Path coefficient (t)	p	Test result
IQ	→	PU	0.266 (2.298)	0.022*	**Supported**	0.391 (3.304)	***	**Supported**
SEQ	→	PU	0.540 (3.254)	0.001**	**Supported**	0.213 (1.676)	0.094	**Supported**
SYQ	→	PU	−0.004 (−0.053)	0.958	Rejected	0.319 (3.103)	0.002**	**Supported**
IQ	→	ESE	0.159 (1.484)	0.138	Rejected	0.088 (1.014)	0.311	Rejected
SEQ	→	ESE	−0.105 (−0.794)	0.427	Rejected	0.135 (1.329)	0.184	Rejected
SYQ	→	ESE	0.070 (0.924)	0.356	Rejected	0.189 (2.080)	0.038*	**Supported**
PU	→	US	0.583 (8.888)	***	**Supported**	0.581 (8.491)	***	**Supported**
ESE	→	US	0.315 (3.805)	***	**Supported**	0.243 (3.006)	0.003**	**Supported**
PU	→	EI	0.207 (2.224)	0.026*	**Supported**	−0.001 (−0.013)	0.989	Rejected
ESE	→	EI	0.942 (6.508)	***	**Supported**	0.706 (6.225)	***	**Supported**
US	→	EI	−0.010 (−0.094)	0.925	Rejected	0.124 (1.425)	0.154	Rejected

*p<0.05, **p<0.01, ***p<0.001

Table 13 Cross-validity analysis of participation in entrepreneurship education difference groups

Division	Unconstrained model	Measurement weights model
χ^2	1186.903	1214.359
X^2 difference	27.456	
P-value of χ^2 difference	0.156	

based on the high and low degree of the relatedness, and the difference between the groups was analyzed.

First, hypothesis that system quality affects perceived usefulness and entrepreneurial self-efficacy is rejected, but it can be confirmed that high related-ness group is supported. This can be interpreted as the acceptance of the system quality of the online entrepreneurship education platform as a way to increase the usefulness and efficacy of the start-up activity. Second, in hypothesis in which per-ceived usefulness affects entrepreneurial intention, the group with low relatedness is supported, but the group with high relatedness is rejected. This can be interpreted that, in the case of the group with low relatedness, they acquire information required for personal start-up activities through online entrepreneurship education platform, as they are insufficient of ability to provide and receive the necessary help for start-up activities in relation to the surrounding people, and that they inspire entrepreneurial intention also through use of online entrepreneurship education platform (Table 12).

5.5.4   Moderating Effects of Participation in Entrepreneurship Education

In order to confirm the moderating effect of participation in entrepreneurship educa-tion, this study verified cross validation proving the metric equivalence of the group never participating in entrepreneurship education (212 persons) and the group having participated in entrepreneurship education group more than once (217 persons). As a result of the cross validity analysis can be explained that the fitness of the model is not deteriorated and the cross validity between comparison groups for participation in entrepreneurship education is secured (Table 13).

As this study could explain that measurement identity between comparison groups for entrepreneurship education participation was secured, and that measurement weight between comparison groups for entrepreneurship education participation was the same, this study confirmed the result with multi comparison analysis by group for entrepreneurship education participation (Table 14).

As this study confirmed that the moderating effect by participation in entrepreneur-ship education between the two groups was significant, this study tried analysis of the difference between the groups as in Table 15 by dividing the sample into two comparative groups according to participation in entrepreneurship education.

Table 14 Multiple comparison analysis between participation in entrepreneurship education difference groups

Division	Unconstrained model	Structural weights model
χ^2	1065.851	1121.503
X^2 difference	55.652	
P-value of χ^2 difference	0.004	

Table 15 Moderating effect analysis result between participation in entrepreneurship education groups

Hypothesis (path)			Non-participation (212)			Participation (217)		
			Path coefficient (t)	p	Test result	Path coefficient (t)	p	Test result
IQ	→	PU	0.375 (3.287)	0.001**	Supported	0.237 (1.993)	0.046*	Supported
SEQ	→	PU	0.359 (2.750)	0.006**	Supported	0.527 (2.841)	0.004**	Supported
SYQ	→	PU	0.138 (1.596)	0.111	Rejected	0.129 (1.436)	0.151	Rejected
IQ	→	ESE	0.151 (1.272)	0.203	Rejected	0.175 (1.665)	0.096	Supported
SEQ	→	ESE	0.032 (0.282)	0.778	Rejected	0.120 (0.839)	0.401	Rejected
SYQ	→	ESE	0.189 (1.852)	0.064	Supported	0.194 (2.323)	0.020*	Supported
PU	→	US	0.433 (7.096)	***	Supported	0.723 (11.285)	***	Supported
ESE	→	US	0.411 (5.728)	***	Supported	0.086 (1.213)	0.225	Rejected
PU	→	EI	−0.058 (−0.699)	0.485	Rejected	0.361 (3.959)	***	Supported
ESE	→	EI	0.720 (6.189)	***	Supported	0.812 (7.989)	***	Supported
US	→	EI	0.333 (2.934)	0.003**	Supported	−0.187 (−2.019)	0.043	Supported

*$p<0.05$, **$p<0.01$, ***$p<0.001$

The results of the comparison between groups upon entrepreneurship education participation show different types of competence and relationship performance. First, in the hypothesis that information quality affects entrepreneurship education self-efficacy, entrepreneurship education participation group was adopted, but entrepreneurship education non-participation group was rejected. This suggests that the entrepreneurship education participation group expects that the content provided by the online entrepreneurship education platform will enhance the efficacy for the entrepreneurship education activity rather than the non-participation group. In other words, it means that entrepreneurship education is the same as previous research results that increase entrepreneurship education self-efficacy. Second, in the hypothesis that entrepreneurial self-efficacy affects user satisfaction, entrepreneurship education participation group was rejected, but it was adopted in entrepreneurship education non-participation group. This can be the results, not that self-efficacy affects system use satisfaction, but that it is caused by verification result between rejected usefulness and user satisfaction that usefulness of contents provided affect user satisfaction. Third, in the hypothesis that perceived usefulness affects the entrepreneurial intention, entrepreneurship education participation group was adopted and entrepreneurship education non-participation group was rejected. This is the same as existing research result that participants usefully receive the contents education effect provided by online entrepreneurship education platform, consequently entrepreneurship education have significant influence on entrepreneurial intention. Fourth, in the hypothesis that user satisfaction affects entrepreneurial intention, the results of adoption are derived from both groups as if there is no difference between groups. However, when the path coefficient (-0.187) and t value (-2.019) of the participating groups are checked, it can be confirmed that they represent negative values.

This is because, in the case of the participating group, significant result value was obtained, but by showing negative values, it can not be said to have a positive effect on the relationship between the two variables. Therefore, it can be interpreted that there is a difference between the two groups. This can be interpreted that the entrepreneurship education participation group rejects and entrepreneurship education non-participation group expects the hypothesis that the user satisfaction on the online entrepreneurship education platform has a positive (+) effect on the entrepreneurial intention.

6 Conclusion

This study empirically verified the causal relationship between the factors affecting the entrepreneurial intention of college students when universities conduct entrepreneurship education using online entrepreneurship education platform, and the study result can be summarized as follows:

First, the quality characteristics (information quality, service quality, system quality) of the online entrepreneurship education platform have a positive (+) effect on the perceived usefulness, and this can be interpreted that the characteristics of online entrepreneurship education platform can be interpreted as having a great influence on the usability of the user as in existing previous studies.

Second, the hypothesis that system quality has positive (+) influence on entrepreneurial self-efficacy among the quality characteristics of online entrepreneurship education platform, but information quality and service quality have positive (+) influence on entrepreneurial self-efficacy, was rejected. This implies that the quality characteristics of the online entrepreneurship education platform do not have a positive effect on increasing the self-efficacy of the user.

In the previous research, the study focused on the relationship between the information system success model and the self-efficacy, perceived usefulness and user satisfaction. However, this study could confirm that the system characteristics of the platform is the factor that affects in increasing the self-efficacy.

Third, it has shown that in the relationship between perceived usefulness, entrepreneurial self-efficacy and user satisfaction perceived usefulness and user satisfaction, which is use characteristics of online entrepreneurship education platform users, have positive (+) influence, respectively. This can be interpreted as a positive effect on user satisfaction on the use of the online entrepreneurship education platform if the use of the online entrepreneurship education platform is useful for start-up activities and enhances user efficacy. The perceived usefulness, which is user utility, has a positive effect on user satisfaction, which is the same as previous research results on the information system success model [3, 38, 39]. This means that when users of the online entrepreneurship education platform feel that the platform is useful for the start-up activity upon using the platform, the user satisfaction for the platform is increased. In addition, the higher the self-efficacy, the higher the user satisfaction.

Fourth, the relationship between perceived usefulness, entrepreneurial self-efficacy, user satisfaction, and entrepreneurial intention, which is a performance factor of online entrepreneurship education platform, has a positive (+) effect on entrepreneurial intention, and the hypothesis that user satisfaction has a positive (+) effect on the entrepreneurial intention was rejected.

This suggests that if online entrepreneurship education platform is useful for start-up activities and that entrepreneurial self-efficacy is high, entrepreneurial intention can be increased, but it means that the entrepreneurial intention is not increased because the user's satisfaction with the platform is high.

Fifth, as a result of verifying the moderating effect between causal variables and outcome variables according to the degree of autonomy, competence, and relationship, which is the basic desire of self-determination, it is not meaningful to analyze differences between groups because autonomy is not ensured cross validity, and it was confirmed that there is a difference between competence and relationship. It has been verified that there is a difference between the groups for the effects of system quality on both competence and relationship on perceived usability and entrepreneurial self-efficacy and perceived usefulness on entrepreneurial intention.

In conclusion, using the online entrepreneurship education platform to encourage the entrepreneurship of college students can increase the entrepreneurial intention by allowing the user to feel that the online entrepreneurship education platform is useful for the start-up activity.

References

1. Anderson, A., Jack, S.: Role typologies for enterprising education: the professional Artisan? J. Small Bus. Enterp. Dev. **15**(2), 259–273 (2008)
2. Bechard, J., Tolohouse, J.: Validation of a didactic model for the analysis of training objectives in entrepreneurship. J. Bus. Ventur. **13**, 317–332 (1998)
3. Bhattacherjee, A.: Understanding information systems continuance. An expectation-confirmation model. MIS Q. **25**(2), 351–370 (2001)
4. Block, Z., Stumpf, A.: Entrepreneurship education research: experience and challenge. In: Sexton, D., Kasarada, J. (eds.) The State of the Art of Entrepreneurship, pp. 17–42 (1992)
5. Boyd, N., Vozikis, G.: The influence of self-efficacy on the development of entrepreneurial intentions and actions, entrepreneurship. Theory Pract. **18**, 63–77 (1994)
6. Chen, C., Greene, P., Crick, A.: Does entrepreneurial self-efficacy distinguish entrepreneurs from managers? J. Bus. Ventur. **13**, 295–316 (1998)
7. Cho, Y.J.: A study on the influence of University entrepreneurial education service quality of entrepreneurial intention: focused on the mediating effects of satisfaction with entrepreneurial education. Asia-Pac. J. Bus. Ventur. Entrepr. **12**(2), 95–103 (2017)
8. Collins, L., Smith, A., Hannon, P.: Applying a synergistic learning approach in entrepreneurship education. Manag. Learn. **37**(3), 335–354 (2006)
9. Davidsson, P.: Researching Entrepreneurship. Springer, Boston, MA (2004)
10. DeLone, W.H., McLean, E.R.: Information system success: the quest for the dependent variable. Inf. Syst. Res. **3**(1), 60–95 (1992)
11. Deci, E.L., Ryan, R.M.: Intrinsic Motivation and Self-Determination in Human Behavior. New York, Plenum Press (1985)
12. Drucker, P.: Innovation and Entrepreneurship. Heinemann, London (1985)
13. Fayolle, A., Gailly, B., Lassas-Clerc, N.: Assessing the impact of entrepreneurship education programs: a new methodology. J. Eur. Ind. Train. **30**(9), 701–720 (2006)
14. Fiet, J.: The pedagogical side of entrepreneurship theory. J. Bus. Ventur. **16**, 101–117 (2001)
15. Fiet, J.: The theoretical side of teaching entrepreneurship. J. Bus. Ventur. **16**, 1–24 (2001)
16. Henry, C., Hill, F., Leitch, C.: The effectiveness of training for new business creation. Int. Small Bus. J. **22**(3), 249–269 (2004)
17. Ireland, R., Hitt, M., Camp, S., Sexton, D.: Integrating entrepreneurship and strategic management actions to create firm wealth. Acad. Manag. **15**(10), 49–63 (2001)
18. Jeon, I.O.: Established business start-up support impact on the youth of business performance. J. Digit. Converg. **10**(11), 103–114 (2012)
19. Jeong, C.G.: Status of University entrepreneurship education for activation of young start-up, entrepreneurship. HRD Rev. **16**(5), (2013)
20. Jo, H.: A study on success model of e-education information system. J. Adv. Inf. Technol. Converg. **13**(6) (2015)
21. Juan, C., Marylene, G.: Understanding e-learning continuance intention in the workplace: a self-determination theory perspective. Comput. Hum. Behav. **24**, 1585–1604 (2008)
22. Kim, A.Y.: Self-determination theory: research and applications in educational settings. Korean J. Educ. Psychol. **24**(3), 583–609 (2010)
23. Kim, C.H., Kang, B.O., Yun, H.B.: An empirical study on the influence of store entrepreneur's start up education and experience before start up to performance of stores. J. Korea Acad.-Ind. Coop. Soc. **14**(3), 1135–1147 (2013)

24. Kim, J.I., Kim, I.H.: The impact of service quality in entrepreneurial education on the self-efficacy, the achievement need and the satisfaction of entrepreneurial education—focusing on the entrepreneurial education of internet shopping mall. Asia-Pac. J. Bus. Ventur. Entrep. **9**(5), 21–31 (2014)
25. Kim, K.H., Kim, Y.T.: Effects of start-up mentoring educational factors on satisfaction with start-up education and start-up intention. Asia-Pac.J. Bus. Ventur. Entrep. **9**(5), 33–41 (2014)
26. Krueger, N., Brazeal, D.: Entrepreneurial potential and potential entrepreneurs. Entrep. Theory Pract. **18**(3), 91–104 (1994)
27. Krueger, N., Reilly, M., Carsrud, A.: Competing models of entrepreneurial intentions. J. Bus. Ventur. **15**(5), 411–432 (2000)
28. Lee, J.Y., Kim, J.R.: Analysis on structural relationships among learners' perceived usefulness, learner satisfaction and related factors in mobile learning in universities. J. Korean Educ. **40**(1), 49–79 (2013)
29. Noel, T.: Effects of entrepreneurial education on intent to open a business. In: Frontiers of Entrepreneurship Research. Babson Conference Proceedings (2001)
30. Park, S.H.: A Study on the Effects of SME Start-up Environment, Entrepreneurial Self-Efficacy and Fear of Business Failure on Entrepreneurial Intention: Focusing on the Moderated Effects of Government's Entrepreneurial Supporting Policy and Mediated Effects of Entrepreneurship. Hoseo University (2017)
31. Peterman, N., Kennedy, J.: Enterprise education: influencing students' perceptions of entrepreneurship. Entrep. Theory Pract. **28**(2), 129–144 (2003)
32. Pitt, L.F., Watson, R.T., Kavan, C.B.: Measuring information system service quality. MIS Q. **21**(2), 195–208 (1995)
33. Ronstadt, R.: The educated entrepreneurs: a new era of entrepreneurial education is beginning. In: Kent, C. (ed.) Entrepreneurship Education, pp. 69–88. Quorum Books, New York (1990)
34. Ryan, R.M., Deci, E.L.: Self-determination theory and the facilitation of intrinsic motivation, social development, and wellbeing. Am. Psychol. **52**, 141–166 (2000)
35. Ryan, R.M., Deci, E.L.: Self-determination theory and the facilitation of intrinsic motivation, social development, and wellbeing. American Psychologist **55**(1), 68–78 (2000)
36. Ryou, K.T.: A Study on Effects of Personal Characteristics and Job Orientation Through Education Factors on the Entrepreneurial Intentions. Choon-Ang University (2012)
37. Saee, J.: A critical evaluation of Australian entrepreneurship education and training. In: Proceedings of the 6th IntEnt Conference in Arnhem, NL (1996)
38. Seddon, P.B., Kiew, M.Y.: A partial test and development of the DeLone & McLean mode of IS success. In: Proceedings of the Fifteenth International Conference on System, pp. 99–110 (1994)
39. Seddon, P.B.: A respectification and extension of the DeLone & McLean mode of IS success. Inf. Syst. Res. **8**(3), 240–253 (1997)
40. Solomon, G., Duffy, S., Tarabishy, A.: The state of entrepreneurship education in the United States: nationwide survey and analysis. Int. J. Entrep. Educ. **1**(1), 65–86 (2002)
41. Urassa, E.A.: Students' entrepreneurial self-efficacy: does the teaching method matter? Educ. Train. **57**(8) (2015)
42. Vesper, K., McMullan, E.: New venture scholarship versus practice: when entrepreneurship academics try the real things as applied research. Technovation **17**(7), 349–358 (1997)
43. Yoo, S.J., Lee, H.Y.: The effect of system characteristic and user characteristic of e-learning systems on learning performance. Korea Service Management Society Proceeding, vol. 11, pp. 225–241 (2007)
44. Young, J.: Entrepreneurship Education and Learning for University Students and Practicing Entrepreneurs. Upstart Publishing, Chicago, IL (1997)
45. Zhao, H., Seibert, S., Hills, G.: The mediating role of self-efficacy in the development of entrepreneurial intentions. J. Appl. Psychol. **90**(6), 1265–1272 (2005)

Author Index

A
Alanni, Russul, 17
Azzawi, Hasseeb, 17

B
Bai, Wenjun, 63
Bao, Siya, 107
Byun, Yungcheol, 177

G
Gerardo, B.D., 177
Ghose, Udayan, 33
Gim, Gwang Yong, 155, 189
Grandhi, Srimannarayana, 75
Gupta, Sharad, 87

H
Hou, Jingyu, 17

I
Im, Eun-tack, 155

J
Jeong, Youngsik, 1

K
Ki, Hong Seok, 189
Kim, Chul, 1

L
Lalis, J.T., 177
Lee, Hoo Ki, 189
Lee, Si-Young, 155
Lee, Sung Taek, 189
Lim, Bum-Taek, 155
Luo, Zhi-Wei, 63

N
Nozaki, Yusuke, 49

O
Oo, Lwin Lwin, 141

P
Park, Namje, 1
Phu, Kay Thinzar, 141

Q
Quan, Changqin, 63

R
Rashmi, 33

S
Sanyal, Sudip, 87
Shin, Soo-Bum, 1
Song, Won-Seok, 155

© Springer Nature Switzerland AG 2019
R. Lee (ed.), *Computer and Information Science*, Studies in Computational
Intelligence791, https://doi.org/10.1007/978-3-319-98693-7

Sung, Younghoon, 1

T
Togawa, Nozomu, 107
Tsai, Jichiang, 123

W
Wibowo, Santoso, 75

X
Xiang, Yong, 17

Y
Yanagisawa, Masao, 107
Yoshikawa, Masaya, 49

Printed in the United States
By Bookmasters